Processing Series No. 4

Series Advisor: Professor B. L. Weiss

SIMOX

Other volumes in this series:

SIMOX

Edited by

Maria J. Anc

Axcelis Technologies Inc.
Beverly, USA

Published by: The Institution of Electrical Engineers, London, United Kingdom

© 2004: The Institution of Electrical Engineers

The Institution of Electrical Engineers,
Michael Faraday House,
Six Hills Way, Stevenage
Herts., SG1 2AY, United Kingdom

British Library Cataloguing in Publication Data

Anc, Maria
    SIMOX. - (EMIS processing series)
    1. Silicon-on-insulator technology
    I. Title
    621.3 ' 1937

    ISBN 0 86341 334 X

Typeset in India by Newgen Imaging Systems (P) Ltd., Chennai
Printed in the UK by MPG Books Limited, Bodmin, Cornwall.

Contents

Contents

Contents

Editor

Maria J. Anc
Axcelis Technologies Inc. 108 Cherry Hill Drive Beverly MA
01915 USA

Maria J. Anc is currently a Staff Scientist at Axcelis Technologies
Inc. in Beverly, MA. She has worked on SIMOX technology process development and materials characterisation when employed
by Ibis Technology Corp. in Danvers, MA, from 1990 to 2001.
Earlier, she worked as a research scientist for the Microsensors
Research Center at the Foxboro Co, Foxboro, MA, and Mas-
sachussetts Microelectronics Center in Westborough, MA. She
has an M.S. degree in electronics from the Technical University of Warsaw and a Ph.D. degree from the Institute of Electron
Technology in Warsaw, Poland. She has more than 40 technical
publications in the solid-state-electronics field and holds 2 patents.
She is a member of the APS, MRS and IEEE.

Authors

Peter L.F. Hemment
School of Electronics and Physical Sciences, University of
Surrey, Guildford, Surrey, GU2 7XH, UK

**Peter L. F. Hemment, B.Sc., Ph.D., D.Sc., Eur.Ing., C.Phys.,
C.Eng., F.Inst.P., FIEE, MIEEE.** Professor Peter Hemment has
more than 30 years' experience in silicon technology with special emphasis on the role of ion beams both for the modification
and analysis of semiconductors and related materials. During the
1980s he initiated and managed projects to develop the process
of ion beam synthesis and is internationally recognised for his
contributions to SOI/SIMOX technology. He focused upon understanding the physics and chemistry of oxygen implanted silicon.
With colleagues he pioneered the use of very high temperature
anneals to achieve complete phase separation leading to the creation of planar structures which transformed SIMOX materials
from being a scientific novelty to an economically viable manufacturing technology. Subsequently he has researched multilayer and
non-continuous SOI structures together with the synthesis of SiGe
and SiGeC layers. He collaborates with academic and industrial
groups and keenly promotes international cooperation.

Harold Hovel
Thomas J. Watson Research Center, IBM, PO Box 218,
Route 134, Yorktown Heights, NY 10598, USA

Harold J. Hovel received his Ph.D. degree from Carnegie Mellon
University in 1968. He joined the IBM T.J. Watson Research Center in 1968 and is currently a Research Staff Member at the IBM
laboratory. His research has been in the area of semiconductor
materials, characterisation, and devices, including III–V compounds and integrated circuit processing, heterojunction devices,
solar cells of many types, Si materials characterisation, and optical
metrology. He has worked on silicon-on-insulator materials since
1990 including both bonded SOI and SIMOX, strained Si, silicon–
germanium, and combinations of these materials. He is the author

of 80 papers, 30 patents, and a number of review articles in both solar energy and SOI materials.

Katsutoshi Izumi
Research Institute for Advanced Science and Technology, Osaka Prefecture University, 1–2 Gakuen-Cho, Sakai 599–8570, Japan

Katsutoshi Izumi received an M.S. degree in electrical engineering from Nagoya Institute of Technology in 1972 and a Ph.D. degree in electronics from Tokyo Institute of Technology in 1988. After joining the Musashino Electrical Communication Laboratory, NTT, in 1972, he performed research mainly on SOI technology for CMOS LSIs. He invented the SIMOX in 1976, and succeeded in fabricating CMOS/SIMOX ICs for the first time in 1978. He is presently working in the field of a new composite semiconductor material for Electron-Photon-Merged devices as a professor in the Research Institute for Advanced Science and Technology at Osaka Prefecture University.

Atsushi Ogura
School of Science and Engineering, Meiji University, 1–1–1 Higashimita, Tama-ku, Kawasaki, Kanagawa 214–8571, Japan

Atsushi Ogura received B.S., M.S., and Ph.D degrees from Waseda University, Tokyo, Japan, in 1982, 1984, and 1991, respectively. He joined Fundamental Research Laboratories, NEC Corporation, Kawasaki, Japan, where he has been engaged in the research on fabrication and evaluation of SOI materials. He was a Visiting Researcher at AT&T Bell Laboratories, NJ, from 1992 to 1994. He left NEC in 2004, and presently he is an associate professor of Meiji University.

Devendra K. Sadana
Thomas J. Watson Research Center, IBM, PO Box 218, Route 134, Yorktown Heights, NY 10598, USA

Devendra K. Sadana received his Ph.D. in Physics from Indian Institute of Technology, New Delhi (India), in 1975. He subsequently joined the University of Oxford, England, as a Research Fellow and worked on ion implanted silicon and gallium arsenide from 1975 to 1979. He worked on similar material systems at

University of California, Berkeley from 1979 to 1983. Dr Sadana became engaged in characterisation of processed silicon, silicon devices and ICs from 1983 to 1987 at Microelectronics Center of North Carolina, Research Triangle Park, NC, and Philips Research Lab in Sunnyvale CA. He joined IBM Research in 1987 and has since been working very closely with IBM's product divisions on a variety of issues pertaining to silicon and SOI technologies. He is presently a Senior Technical Staff Member and heads a group of material scientists, physicists and electrical engineers to develop materials and processes for IBM's future generation CMOS technologies. He holds 33 patents, has published over 120 technical articles in refereed journals and proceedings, has written six book chapters, and has given numerous invited talks and industrial courses in the last three decades of his scientific career. He received IBM's corporate award on SOI CMOS technology in 2001.

Abbreviations

Acronym	Chapter	Meaning
AC	5	alternating current
ADM	6	addressable memory
AES	2, 4	Auger electron spectroscopy
AFM	6	atomic force microscopy
ATM	2	asynchronous transfer mode
BOX	1, 3–6	buried oxide
BSOI	6	bonded silicon-on-insulator
CAD	3	computer aided design
CAGR	1	compound annual growth rate
CMOS	1–3, 5, 6	complementary metal-oxide semiconductor
CMP	1	chemical-mechanical polishing
COS	5	corona oxide semiconductor
DC	5	direct current
DIBL	5	drain induced barrier lowering
DRAM	4, 6	dynamic random-access memory
FD-MOS	3	fully depleted metal oxide semiconductor
FET	1, 5, 6	field-effect transistor
FZ	2	float-zone
GOI	5	gate oxide integrity
GXR	5	glancing X-ray reflection
HBT	3	heterojunction bipolar transistor
HTA	3	high temperature anneal
IBS	3	ion beam synthesis
IC	1–4, 6	integrated circuit
IR	5	infrared
ITOX	2–6	internal thermal oxidation
ITRS	3	International Technology Roadmap for Semiconductors
LD	6	low-dose
LII	3, 4	light ion implantation
LSI	1, 2	large scale integration

MEMS	3	microelectromechanical systems
MLD	1, 4, 6	modified low-dose
MOS	1, 3, 5	metal oxide semiconductor
MOSFET	2, 5, 6	metal oxide semiconductor field-effect transistor
MSI	1	million square inch
PC	5	photoconductivity
PD-MOS	3	partially depleted metal oxide semiconductor
PIG	2, 3	Penning ionisation gauge
PIII	3	plasma immersion ion implantation
RF	1, 6	radio frequency
RIE	5	reactive ion etching
RMS	5, 6	root mean square
RT	4, 6	room temperature
SE	5	spectroscopic ellipsometry
SEM	2, 6	scanning electron microscopy
SGV	5	surface generation velocity
SIA	5	Semiconductor Industry Association
SIMOX	1–6	separation by implanted oxygen
SIMS	3–5	secondary ion mass spectroscopy
SOC	4, 6	system-on-chip
SOI	1–6	silicon-on-insulator
SON	4	silicon-on-nothing
SOS	2, 3, 6	silicon-on-sapphire
SPIMOX	3	separation by plasma implanted oxygen
SRAM	1, 2, 6	static random-access memory
SRV	5	surface recombination velocity
SSL	3	solid solubility limit
SSL	5	subthreshold slope
TDI	3	total device isolation
TEM	4–6	transmission electron micrograph (or microscopy)
TMAH		tetra-methyl-ammonium hydroxide
ULSI	4	ultra-large-scale integration
UV	5	ultraviolet
VLSI	2	very-large-scale integration
XRD	5	X-ray diffraction
XTEM	2, 3, 6	cross-sectional transmission electron microscopy

Chapter 1

Introduction

M.J. Anc

SIMOX, separation-by-implanted-oxygen, is a method of fabrication of silicon-on-insulator (SOI) structures and wafers by implanting high doses of oxygen and annealing at high temperature. SIMOX distinguishes itself among the SOI fabrication techniques by excellent uniformity of thin film layers, low defect density, and feasibility of patterned SOI wafers. Suitable for large area wafers and volume production, SIMOX SOI has shown the capability to support present LSI applications [1].

For many years, considerations have been given to whether SOI may become a mainstream technology or will only remain in niche applications. Finally, an unavoidable need to solve the problem of power consumption and boost performance of scaled down circuits has led to the recognition of the benefits of SOI. As shown in FIGURE 1.1, the International Technology Roadmap for Semiconductors finds SOI wafers as a starting substrate solution for the upcoming hp65 technology node [2]. The pre-production qualification stage is seen to come in year 2006, followed by developments of next generation SOI substrates enhanced with high resistivity silicon and strained layers.

Concurrently, a steady growth of demand for SOI wafers is forecasted by market research [3]. The increase in demand for SOI wafers is expected to arise first from the demand in high-end applications such as data processing, communications and consumer electronics. With prevalence of $0.13\,\mu m$ and $0.09\,\mu m$ processes and reduced cost per chip in 300 mm wafer generation, SOI is expected to expand then to mid-range and low-end segment applications. The CAGR (compound annual growth rate) for SOI wafers is predicted to record 41 percent from 2002 to 2008, and reach 254 MSI (million square inch) in 2008. At the same time, the demand for ultrathin/thin film wafers is expected to increase at an even faster rate and account for 81 percent of the total, or reach 203 MSI, which gives very favourable prospects for thin film SOI technologies [3].

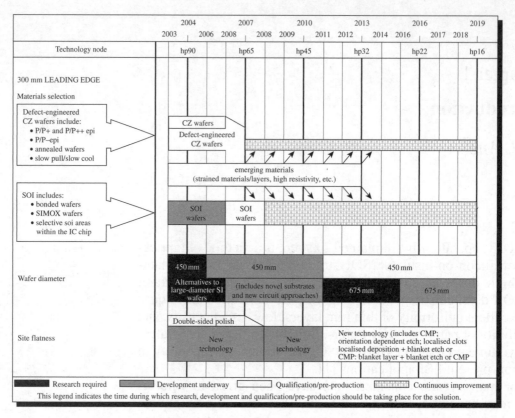

FIGURE 1.1   Starting material solutions for future technology nodes as outlined in the 2003 edition of International Technology Roadmap for Semiconductors [2].

SIMOX is one of few SOI technologies capable of supporting thin film SOI applications in volume production. It was used to demonstrate the first CMOS circuit built in SOI and reported in 1978 [4], and since then has overcome numerous conceptual and technological limitations to grow to a manufacturable fabrication process [5]. Over the years, the technical progress in SIMOX has been reported by groups of researchers and technologists in various scientific journals and conference proceedings. This book is the first effort to compile broad knowledge of SIMOX technology scattered in the technical literature or not yet published into a concise volume. It shows challenges that SIMOX technology has had to face and crucial milestones that have advanced it. The book content includes theoretical background underlying formation of SIMOX buried oxide, fabrication processes, fundamental material characteristics and characterisation methodology, and describes crucial advancements and evolution of material fabrication methods from the experimental research stage to production-worthy processes proven to support advanced logic SOI products manufactured at IBM. The book consists of a sequence of chapters, each dedicated to a specific aspect of SIMOX technology.

Chapter 2 introduces the reader to the concept of SOI technology and SIMOX. Prof. Katsutoshi Izumi provides a review of major challenges and technological achievements that made a dramatic impact on the growth of SOI and SIMOX technologies from the early 1970s to late 1990s. To realise the long-term vision of future technology with superior device/circuit performance and low power consumption, novel devices and circuits needed to be conceived (i), a new generation of equipment capable of implanting very large doses and operating at elevated temperatures needed to be developed (ii), and manufacturability of SIMOX wafer processing had to be improved (iii). These three aims have been providing the driving force for developments in SIMOX technology for the past decades, and with an updated focus on new requirements are likely to continue to play an important role in the future.

In Chapter 3 Prof. Peter L.F. Hemment provides a detailed description of the fundamental processes underlying formation of SIMOX buried oxide. Referring to the earliest published work on implanted oxides and the persistent continuation of efforts to understand the mechanisms of formation of high quality buried insulating layers in a predictable manner, the author explains the phenomenology, physics and chemistry of oxide synthesis by ion implantation and annealing. While giving an in-depth scientific background on mechanisms of buried oxide formation, consideration is also given to the fabrication processes and material and equipment engineering options. Chapter 3 is concluded with an intriguing philosophical question on the evolution of SOI technology into the next generations of semiconductor technology and the future role of ion implantation in new emerging concepts.

Chapter 4 elaborates on an unconventional process approach, and the capability of formation of buried oxide with the implanted doses outside the dose-energy (process window) by controlling precipitation of oxygen. Dr. Atsushi Ogura focuses on the design of annealing processes with selected ambient conditions and temperature ramp rates, and demonstrates various multiple SIMOX layers formed under fine-tuned process conditions. Demonstration of extreme effects of process design on formation of buried layers may challenge and encourage further exploration of the potential and limits of flexibility of SIMOX processes and structures.

In Chapter 5 Dr. Harold Hovel provides an in-depth understanding of SOI characterisation methods, with special emphasis on adequate measurements of properties of starting SIMOX SOI substrates that are important for material quality and reproducibility in advanced applications. Reflectometry, ellipsometry, X-ray reflection, pseudo-FETs including Hg based, MOS capacitors, BOX capacitors, gate oxide capacitors, recombination and generation lifetime, Hall measurements and contactless corona discharge are among the techniques used and/or developed by the author at IBM to characterise SOI substrates. The significance of characterisation efforts in development of manufacturable technology is demonstrated by explanation of the principles of these techniques with application to measured material properties, interpretation of the results with respect to process conditions, and material defects and their potential impact on devices. Spectroscopic ellipsometry and pseudo-FETs, particularly in the Hall FET configuration, methods exploited extensively by the author in SOI substrate characterisation, are applicable down to sub-10 nm thicknesses and can be expected to play an invaluable role in continued material development and quality control.

In Chapter 6 Dr. Devendra Sadana describes the evolution of material fabrication processes from the perspective of development of a robust process, competitive with other SOI material technologies, capable of supporting manufacture of advanced SOI products. The modern modified low-dose (MLD) process is shown to achieve yield equivalent to bonded SOI in functional products including microprocessors, SRAM memories and high frequency RF circuits fabricated utilising IBM's 0.18 µm and 0.13 µm CMOS technologies. Manufacturable and attractive from the point of view of cost of ownership, applicable in a selective pattern on the wafer and scalable to sub-0.1µm geometries, the patterned MLD SIMOX is envisaged to play a leading role in future technology nodes when increase in performance will extend from chip-level to system-on-chip level.

As technology advances, new applications may emerge and new materials, processes and characterisation methods may have to be developed. Considering trends and forecasts in semiconductor technology [2,3] it can be anticipated that SOI techniques will play a significant role in the evolving technology nodes. This book will provide a unique (at this time) source of knowledge about SIMOX and SOI. In addition to the topics directly pertaining to SIMOX, the content of the book offers a wealth of information on ion implantation, thermal processing in extreme conditions, defect formation, characterisation techniques, applications and future technology trends. The exemplary history of growth of SIMOX, with the necessity to overcome severe technological obstacles and limitations to turn it into a successful technology, will hopefully inspire the technical community to explore further the possibilities and potential of SIMOX processes, and stimulate novel technology developments.

As editor of this book, I would like to express my appreciation for the privilege of working on it, and to thank all the authors, Profs. Katsutoshi Izumi and Peter Hemment, and Drs. Atsushi Ogura, Harold Hovel and Devendra Sadana, for their commitment to writing this book and effort in preparation of the manuscript, and especially for their willingness to share their invaluable experience and contributions on the science and technology of SIMOX SOI with a broad audience of readers.

REFERENCES

[1] G. Shahidi et al. [*Proc. 1999 IEEE Int. SOI Conf.* (1999), p.1–4, ISBN 0-7803-5456-7]

[2] [*The International Technology Roadmap* (SIA, 2003 edition)] http://public.itrs.net

[3] Gartner Dataquest [Focus report on SOI (November 2003)]

[4] K. Izumi, M. Doken, H. Ariyoshi [*Electron. Lett. (UK)* vol.14 (1978) p.593–4]

[5] R. Hannon et al. [*VLSI Tech. Dig.* (2000) p.66–8]

Chapter 2

Overview of SIMOX technology: historical perspective

K. Izumi

2.1 INTRODUCTION

In 1966, Watanabe and Tooi [1] first reported on the formation of silicon oxide by oxygen-ion implantation into silicon. Their work, however, has not been followed by additional SOI (silicon-on-insulator) studies.

In 1978, Izumi, Doken and Ariyoshi succeeded in fabricating a 19-stage CMOS (complementary metal-oxide semiconductor) ring oscillator using a buried SiO_2 layer formed by oxygen-ion ($^{16}O^+$) implantation into silicon. Izumi named the new SOI technology "SIMOX," which is short for separation by implanted oxygen [2]. Since then, Izumi and his research group have been constantly committed to the study of SIMOX technology.

SIMOX was originally conceived in order to solve the aluminium auto-doping and crystalline defect problems in SOS (silicon-on-sapphire). There were three key points in SIMOX technology. First, the buried oxide layer must be formed with adequate dielectric isolation characteristics. Second, the top silicon layer above the buried oxide layer must maintain sufficient monocrystallinity. Third, the thermally oxidised silicon layer, which covers the mesa-type island of a MOSFET (metal-oxide semiconductor field-effect transistor), and the buried oxide layer must be combined with sufficient continuation at the bottom edge of the mesa-type island.

Basically, SIMOX technology satisfied these three key points relatively early; however, it still faced other serious difficulties. The synthesis of SiO_2 by oxygen-ion implantation into silicon requires a huge oxygen dose. Simple calculations indicate that the synthesis of stoichiometrical SiO_2 requires high-density oxygen atoms of 4.48×10^{22} cm^{-3}. Taking an acceleration energy of $^{16}O^+$ at 150 keV, the necessary oxygen dose is in the order of 10^{18} cm^{-2}. In other words, the oxygen dose is 100 to 1,000 times higher than

7

the impurity-ion dose employed in conventional semiconductor fabrication processes. This was beyond the capabilities of a conventional ion implanter in the 1970s. A conventional ion implanter of the late 1970s with an $^{16}O^+$ beam current of about $100\,\mu A$ required approximately 65 hours to form the buried SiO_2 layer by $^{16}O^+$ implantation into a 100 mm silicon wafer. Only after the joint development of a 100-mA-class oxygen-ion implanter (NV-200) by Eaton Corporation and NTT in 1986 was SIMOX considered to be one of the most practical and attractive SOI technologies. In addition, the development of the low-dose SIMOX wafer in 1990 and the ITOX (internal thermal oxidation) process in 1994 by Izumi and his research group put the SIMOX technology into the dominant position among SOI technologies, and accelerated its use in practical applications.

In the following, SIMOX technology is overviewed from its origins to its current state of practical use, by classifying its development into three stages.

2.2 THE FIRST STAGE

Izumi, Doken and Ariyoshi succeeded in fabricating a 19-stage CMOS ring oscillator using the early SIMOX wafer in 1978, employing a $5\,\mu m$ design rule. Furthermore, in 1982, Izumi et al. reported on a fabricated 1 kb CMOS SRAM (static random-access memory) using the SIMOX wafer and a $2\,\mu m$ design rule. The success of the 1 kb CMOS SRAM was the gateway for SIMOX technology to be used in LSI/SOI applications.

2.2.1 Concept of SIMOX technology

In 1976, Izumi started the SIMOX study with a desire to develop a thoroughly isolated MOSFET by completely wrapping the MOSFET in SiO_2. If realised, this type of MOSFET was thought to offer possibilities of solving the auto-doping of aluminium and other associated problems encountered in SOS. For this reason, nitrogen-ion implantation into silicon was not adopted from the start.

The concept of SIMOX technology is shown in FIGURE 2.1. FIGURE 2.1(a) shows the formation of a buried SiO_2 layer by oxygen-ion implantation into silicon. FIGURE 2.1(b) shows a cross-sectional view of a MOSFET/SIMOX using the buried SiO_2 as an insulator. There were three key points in SIMOX technology. First, the buried oxide layer must be formed with adequate dielectric isolation characteristics. Second, the top silicon layer

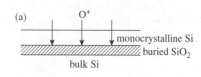

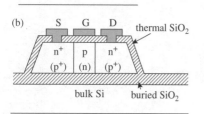

FIGURE 2.1   Concept of SIMOX technology. (a) Formation of a buried SiO_2 layer by oxygen-ion implantation into silicon. (b) Cross-sectional view of a MOSFET/SIMOX using the buried SiO_2 as an insulator.

above the buried oxide layer must maintain sufficient monocrystallinity. Third, the thermally oxidised SiO_2 layer, which covers the mesa-type island of a MOSFET, and the buried oxide layer must be combined with sufficient continuation at the bottom edge of the mesa-type islands as shown in FIGURE 2.1(b).

2.2.2 Extraction of oxygen-ion beam

The implanter used in the oxygen-ion implantation study in 1976 was an Extrion 200-20A that was popular at that time. The ion source of this implanter was a cold cathode PIG (Penning ionisation gauge), as shown in FIGURE 2.2. As such, it was thought to be of a construction that would also amply withstand active gases. For the sake of caution, however, CO_2, of relatively low activity, was adopted at the outset as the source gas. FIGURE 2.3 shows the ion beam spectra at the end station for an acceleration energy of 150 keV. Only 30 μA for an $^{16}O^+$ beam current was attained. In addition, carbon was formed by the breakdown of CO_2 and deposited on the inner walls of the ion source. The deposited carbon degraded the electrode insulation. To overcome this problem, the source gas was changed from CO_2 to O_2. This resulted in an increase of the $^{16}O^+$ beam current from 30 μA to 50 μA, but a new problem then surfaced. This was an extremely rapid diminution in the life of the ion source aperture. This phenomenon was hardly surprising because, as the aperture was made of a volcanic lava consisting mainly of carbon, it wore out easily through reaction with the active $^{16}O^+$ beam. Thus, the aperture problem required immediate solution. Conductive substances that would withstand high temperature and not easily be magnetised were considered. Finally, titanium was selected. The use of an aperture made of titanium increased the $^{16}O^+$ beam current from 50 μA to 100 μA and extended the aperture life time from 5 to 100 hours. As a result, a stable $^{16}O^+$ beam current – relatively large by contemporary standards – became available from the Extrion 200-20A. This opened the way for oxygen-ion implantation experiments.

2.2.3 Formation of buried oxide

It was necessary from the start to ensure that the SiO_2 layer could be formed by oxygen-ion implantation into silicon, and that this SiO_2 layer would have adequate electrical insulation characteristics. FIGURE 2.4 shows the infra-red spectra produced by implanting $^{16}O^+$ with a 1×10^{18} cm^{-2} dose into silicon at an acceleration energy of 32 keV [3]. As can be seen in FIGURE 2.4, $^{16}O^+$ implantation and annealing of at least 900°C were confirmed

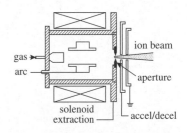

FIGURE 2.2 Cross-sectional view of the cold cathode PIG mounted on Extrion 200-20A.

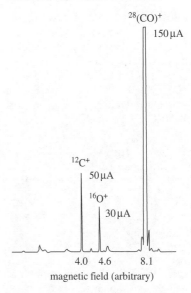

FIGURE 2.3 Ion beam spectra in case of using CO_2 as a source gas and lava aperture.

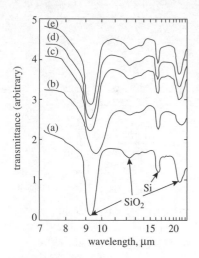

FIGURE 2.4 Infra-red spectra on the oxygen-implanted silicon layers. (a) Thermally grown SiO$_2$ as a control specimen; (b) as-implanted; (c) annealed at 900°C for 1 h in N$_2$; (d) annealed at 900°C for 9 h in N$_2$; (e) annealed at 1150°C for 1 h in N$_2$.

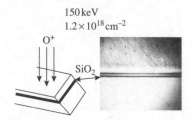

FIGURE 2.5 Photomicrograph of an oxygen-implanted silicon layer after annealing at 1150°C for 2 h in N$_2$, followed by polishing to a bevelled angle.

to be capable of forming an SiO$_2$ layer equal to thermally grown SiO$_2$ from the infra-red spectroscopy analysis standpoint. The electrical and physical characteristics of this SiO$_2$ layer were examined. It was found that the dielectric strength, dielectric constant and refractive constant of the SiO$_2$ annealed at 1150°C for 2 hours were 8.5×10^6 V cm^{-1}, 3.83 and 1.46, respectively. It was evident that the electrical insulation characteristics of the SiO$_2$ layer formed by ^{16}O$^+$ implantation were sufficient for MOSFET application.

For SOI, the SiO$_2$ layer formed by ^{16}O$^+$ implantation must be buried while retaining the monocrystallinity of the top silicon layer. Assuming that the concentration profile of the oxygen atoms in silicon is a Gaussian distribution, ^{16}O$^+$ was implanted into a silicon substrate under the conditions of an implantation energy of 150 keV and a dose of 1.2×10^{18} cm^{-2}; these conditions were determined using the table given by Smith [4] in order to locate the projected range at a depth of 0.37 μm from the surface of a silicon substrate and to give sufficient (dose of) oxygen atoms to form SiO$_2$ stoichiometrically.

FIGURE 2.5 shows a photomicrograph taken in 1976 of the specimen after annealing at 1150°C for 2 hours in N$_2$, followed by polishing to a bevelled angle [5]. The formation of the buried SiO$_2$ layer is obvious. This specimen was also analysed by AES (Auger electron spectroscopy) and the results are shown in FIGURE 2.6 [2]. A considerable quantity of oxygen and carbon can be seen in FIGURE 2.6 at the surface portion of the specimen. The projected range of the oxygen atoms is located at a depth of 0.38 μm, which corresponds to the depth estimated from Smith's table. The thickness of the buried SiO$_2$ layer was found to be 0.21 μm, which was also measured by SEM (scanning electron microscopy) after cleaving the specimen and etching it in buffered HF.

The next question is whether the silicon of this surface layer is monocrystalline. FIGURE 2.7 shows the electron-beam diffraction pattern of this specimen whose surface was slightly etched off to scrape the oxygen- and carbon-contaminated surface portion. The Kikuchi bands and lines are clearly visible, attesting to the monocrystalline structure of the specimen's surface. In this way, in 1976, the basic requirements were satisfied to realise MOSFET/SOI. Since that time, the technology has been called "SIMOX" [2].

2.2.4 Realisation of CMOS IC

Subsequently, the technologies for epitaxial silicon growth on SIMOX substrates, the formation of a mesa-type island, etc., were developed to fabricate a pMOSFET. FIGURE 2.8(a) shows the Ids-Vds characteristics of the pMOSFET fabricated on a trial basis in 1977 [5]. The active layer and gate oxide of the fabricated pMOSFET were 0.5 μm and 70 nm thick, respectively. The Ids-Vds characteristics shown in FIGURE 2.8(a) were memorable in that this was the first time the operating reality of a MOSFET/SIMOX was demonstrated. For comparison, a pMOSFET/SOS was fabricated simultaneously by the same process and its Ids-Vds characteristics are shown in FIGURE 2.8(b). A comparison revealed that the electrical characteristics of the pMOSFET/SIMOX were superior to those of the pMOSFET/SOS. For instance, when Vg = Vds = −10 V, the Ids of the pMOSFET/SIMOX is 11.2 mA and that of the pMOSFET/SOS is 7.3 mA.

Encouraged by these achievements, the trial efforts next aimed at the fabrication of a CMOS/SIMOX IC. FIGURE 2.9 is a photomicrograph of a 19-stage CMOS/SIMOX ring oscillator fabricated using a 5-μm design rule. The propagation delay time of this oscillator was a fast 0.96 ns/stage for an applied voltage of 5 V. This success in CMOS/SIMOX IC took place in 1978 and these achievements were published for the first time in the same year as the reports on SIMOX technology [2].

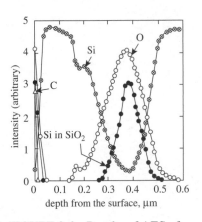

FIGURE 2.6   Results of AES of the specimen implanted with an $^{16}O^+$ dose of 1.2×10^{18} cm^{-2} at 150 keV, after annealing at 1150°C for 2 h in N_2.

2.2.5   Potential for LSI

Based on the successful fabrication of the CMOS/SIMOX IC, a 1 kb CMOS/SIMOX SRAM was fabricated by a 2-μm design rule in 1982 on a trial basis, using MOSFETs of cross-sectional structure shown in FIGURE 2.10. FIGURE 2.11 shows a photomicrograph of the trial chip [6]. The chip-select access time of this

FIGURE 2.7   Electron-beam diffraction pattern of the specimen described in FIGURE 2.6 after slightly scraping the surface portion by etching.

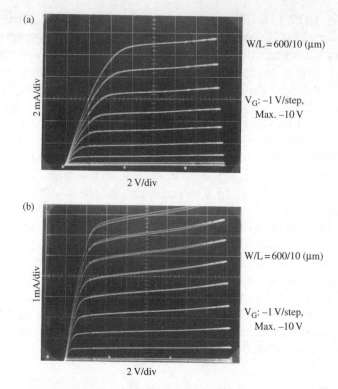

FIGURE 2.8   Comparison of the Ids-Vds characteristics of pMOSFETs.
(a) pMOSFET/SIMOX; (b) pMOSFET/SOS.

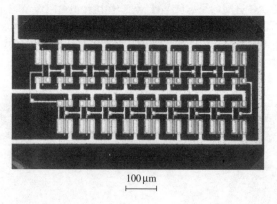

FIGURE 2.9   Photomicrograph of a 19-stage CMOS/SIMOX ring oscillator
fabricated by a 5-μm design rule.

SRAM was 12 ns for an applied voltage of 5 V. The success of the
1 kb CMOS SRAM demonstrated that SIMOX technology had a
high potential for LSI applications. However, two serious problems in SIMOX technology remained unsolved at this stage: the
unpractically long implantation time for the formation of the buried oxide layer and the high dislocation density in the top silicon
layer.

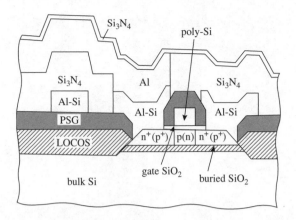

FIGURE 2.10 Cross-sectional structure of the MOSFET/SIMOX used for the 1 kb CMOS SRAM.

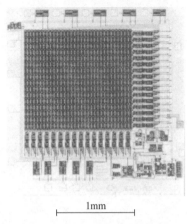

FIGURE 2.11 Photomicrograph of a 1 kb CMOS/SIMOX SRAM fabricated by a 2-μm design rule.

2.3 THE SECOND STAGE

In 1986, a 100-mA-class oxygen-ion implanter, NV-200, was jointly developed by Eaton Corporation and NTT to solve the unpractically long implantation time for the formation of the buried SiO_2 layer. In 1990, Izumi and his research group succeeded in lowering the dislocation density from the conventional $10^9 \, cm^{-2}$ to less than $100 \, cm^{-2}$ by reducing the $^{16}O^+$ dose from the conventional $2 \times 10^{18} \, cm^{-2}$ (hereafter referred to as a "high-dose SIMOX") to $4 \times 10^{17} \, cm^{-2}$ (hereafter referred to as a "low-dose SIMOX") when the acceleration energy was approximately 180 keV. The development of the NV-200 and the drastic reduction of the dislocation density in the top silicon layer greatly activated the study of SIMOX technology for LSI/SOI applications.

2.3.1 Development of high-current oxygen implanter

SIMOX technology was, until the mid-1980s, nearly unanimously judged unfit for practical application. This verdict was understandable in a sense because SIMOX requires $^{16}O^+$ implantation at an extremely high dose in the order of $10^{18} \, cm^{-2}$. This is 100–1,000 times the dose required for impurity doping in conventional semiconductor fabrication processes.

Up to this point, a conventional ion implanter with $^{16}O^+$ beam current of about 100 μA was used for the formation of the buried SiO_2 layer. This method required approximately 65 hours to perform $^{16}O^+$ implantation with a dose of $1.2 \times 10^{18} \, cm^{-2}$ into a 100-mm-silicon wafer. In 1986, a 100-mA-class oxygen-ion

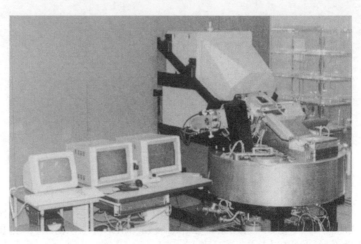

FIGURE 2.12 External view of the 100-mA-class oxygen implanter, NV-200.

implanter, NV-200, was jointly developed by Eaton Corporation and NTT to solve the unpractically long implantation time [7].

An external view of the NV-200 is shown in FIGURE 2.12. The NV-200 took only 5 hr per 100-mm-silicon wafer to perform the necessary $^{16}O^+$ implantation. The completion of this implanter renewed interest in SIMOX technology in many fields, leading to a sharp increase in the number of SIMOX researchers.

2.3.2 Development of low-dose SIMOX wafer

A conventional SIMOX wafer (high-dose SIMOX) was produced by $^{16}O^+$ implantation into silicon with a dose of approximately 2×10^{18} cm^{-2} at an acceleration energy of 180 keV, followed by high-temperature annealing. The resulting thicknesses of the top silicon layer and the buried SiO_2 layer were 0.2 μm and about 0.5 μm, respectively. In this case, the top silicon layer had typical high-density dislocations of approximately 10^9 cm^{-2}. In 1990, Izumi and his research group succeeded in reducing the oxygen dose to drastically lower the dislocation density, as shown in FIGURE 2.13 [8]. This method lowers the dislocation density from 10^9 cm^{-2} to the order of 10^6 to 100 cm^{-2} when the oxygen dose is less than 1.4×10^{18} cm^{-2} at an acceleration energy of 180 keV and when the annealing temperature is higher than 1300°C.

FIGURE 2.14 shows X-TEM (cross-sectional transmission electron microscopy) photomicrographs of the specimens taking the oxygen dose as a parameter [9]. The top photomicrographs are from specimens as-implanted or before annealing, whereas the bottom ones are from specimens after annealing at 1350°C for 4 hours. Either discontinuous buried SiO_2 layers in silicon or silicon particles in the buried SiO_2 are seen in the annealed

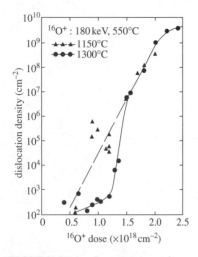

FIGURE 2.13 Dependence of dislocation density on oxygen dose.

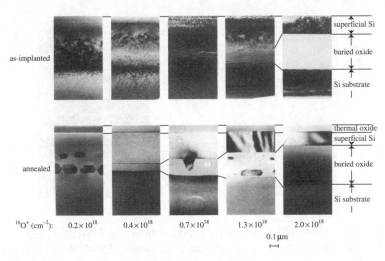

superficial Si

buried oxide

Si substrate

thermal oxide
superficial Si

buried oxide

Si substrate

as-implanted

annealed

$^{16}O^+$ (cm^{-2}): 0.2×10^{18} 0.4×10^{18} 0.7×10^{18} 1.3×10^{18} 2.0×10^{18}

0.1 μm

FIGURE 2.14   Comparison of X-TEM photomicrographs of specimens implanted with oxygen doses ranging from 0.2 to 2.0×10^{18} cm^{-2} at an acceleration energy of 180 keV, before (top) and after (bottom) high-temperature annealing at 1350°C for 4 h.

specimens depending on the oxygen dose. When the oxygen dose is 4×10^{17} cm^{-2}, however, the buried SiO_2 layer is continuous without any silicon particles. In addition, the dislocation density is as low as 100 cm^{-2}. This dose is referred to as a "dose-window," whereas the wafer is called a "low-dose SIMOX". The drastic reduction of the dislocation density in the top silicon layer activated the study of SIMOX technology for LSI/SOI applications.

2.4   THE THIRD STAGE

In 1994, Izumi and his research group developed a high-temperature oxidation technique referred to as ITOX (internal thermal oxidation), which decisively improved the quality of the buried SiO_2 layer and reduced the micro-roughness of the top silicon layer. The usefulness of the ITOX-SIMOX wafers was demonstrated by applying them to a high-speed 300 kb gate array [10] and to a high-performance 8×8 ATM switch [11], which was a full-custom VLSI (circuit) composed of 2 million transistors, using the 0.25-μm design rule.

2.4.1   Invention of ITOX process

Izumi and his research group invented a high-temperature oxidation technique to improve the quality of the buried SiO_2 layer and to reduce the micro-roughness of the top silicon layer in 1994.

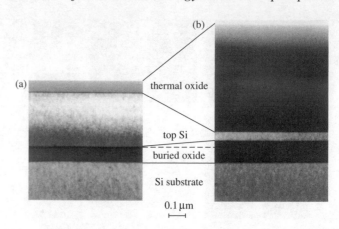

FIGURE 2.15   X-TEM photomicrographs of low-dose SIMOX wafers before (a) and after (b) high-temperature oxidation at 1350°C (ITOX process).

FIGURE 2.15(a) shows an X-TEM photomicrograph of a low-dose SIMOX wafer implanted with an oxygen dose of $4 \times 10^{17} \, \mathrm{cm}^{-2}$ at an acceleration energy of 180 keV, followed by annealing at 1350°C. Additional high-temperature oxidation at 1350°C causes the growth not only of superficial thermal oxide, but also of internal thermal oxide on the original buried SiO_2, as shown in FIGURE 2.15(b) [12]. The high-temperature oxidation technique is referred to as ITOX. The undulation at the top silicon/buried SiO_2 interface is less than ± one lattice.

FIGURE 2.16 shows a normalised oxygen concentration in the top silicon layer of an ITOX-SIMOX wafer, taking the oxygen concentration of an FZ (float-zone refined) silicon wafer as a reference [10]. There is no difference in residual oxygen concentration between an ITOX-SIMOX wafer and an FZ silicon wafer. FIGURE 2.17 compares the breakdown voltage of the buried SiO_2 layer before and after the ITOX process, taking a thermally oxidised SiO_2 layer as a reference. As shown in FIGURE 2.17, the breakdown voltage of the buried SiO_2 layer is much improved by the ITOX process.

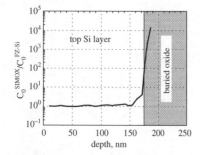

FIGURE 2.16   Oxygen concentration in the top silicon layer of an ITOX-SIMOX wafer normalised by the oxygen concentration of an FZ silicon wafer (courtesy of Komatsu Electronics Metals Co., Ltd.).

2.4.2   Application to VLSI

The thickness dispersions of the top silicon layer are as small as or less than 2.0 nm and independent of the layer thickness across a 200 mm ITOX-SIMOX wafer. The excellent thickness uniformities are basically the result of the use of ion-implantation technology.

In 1996, Ino et al. [10] reported a high-speed 300 kb gate array VLSI for telecommunication use, which contains 3.2 million transistors, fabricated using the ITOX-SIMOX wafer and a 0.25-μm

16

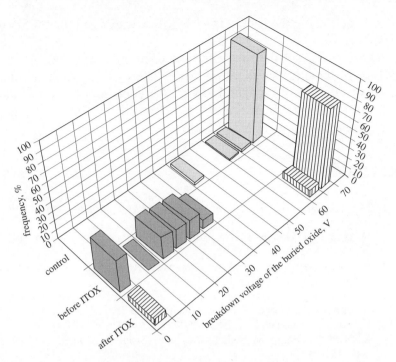

FIGURE 2.17   Improvement in the breakdown voltage of the buried SiO_2 layer by the ITOX process.

FIGURE 2.18   Photomicrograph of a 300 kb gate array VLSI fabricated using an ITOX-SIMOX wafer and a 0.25-μm design rule. Chip size: $10 \times 10\,mm^2$. Transistors used: 3.2 million.

design rule. FIGURE 2.18 shows the photomicrograph of the VLSI chip. In addition, Ohtomo et al. [11] succeeded in fabricating a high-performance 8×8 ATM (asynchronous transfer mode) switch in 1997, which was a full-custom VLSI composed of 2 million transistors, using the same ITOX-SIMOX wafer and the 0.25-μm design rule.

These successes demonstrated the usefulness of SIMOX technology for practical application to VLSIs, especially in the field of high-speed, low-voltage, and low-power consumption. Currently, 300 mm ITOX-SIMOX wafers are available, which is greatly expediting the use of SIMOX technology.

REFERENCES

[1] M. Watanabe, A. Tooi [*Jpn. J. Appl. Phys. (Japan)* vol.5 (1966) p.737–8]

[2] K. Izumi, M. Doken, H. Ariyoshi [*Electron. Lett. (UK)* vol.14 (1978) p.593–4]

[3] K. Izumi, M. Doken, H. Ariyoshi [*Jpn. J. Appl. Phys. (Japan)* vol.19, supplement 19-1 (1980) p.151–4]

[4] G. Dearnaley, J.H. Freeman, R.S. Nelson, J. Stephen [*Ion Implantation* (North-Holland, London, 1973) Appendix 3, p.766]

[5] K. Izumi [*Vacuum (UK)* vol.42 (1991) p.333–40]

[6] K. Izumi, Y. Omura, M. Ishikawa, E. Sano [*Symp. VLSI Tech. Dig.* (1982) p.10–11]

[7] J.P. Ruffell, D.H. Douglas-Hamilton, R.E. Kaim, K. Izumi [*Proc. 6th Int. Conf. Ion Implant. Tech.* (1986) p.229–34]

[8] S. Nakashima, K. Izumi [*Electron. Lett. (UK)* vol.26 (1990) p.1647–9]

[9] S. Nakashima, K. Izumi [*Nucl. Instrum. Methods Phys. Res. B (Netherlands)* vol.55 (1991) p.847–51]

[10] M. Ino et al. [*IEEE ISSCC Digest* (1996) p.86–7]

[11] Y. Ohtomo et al. [*IEEE ISSCC Digest* (1997) p.154–5]

[12] S. Nakashima et al. [*Proc. IEEE Int. SOI Conf.* (1994) p.71–2]

Chapter 3

Fundamental processes in SIMOX layer formation: ion implantation and oxide growth

P.L.F. Hemment

> "God gave silicon a surface oxide,
> Man gave silicon a buried oxide"
> Jean-Pierre Colinge [1]

3.1   INTRODUCTION

Ion implantation is the term used to denote the injection of impurity atoms in the form of fast ions into a target for the purpose of modifying the properties of the target. When this results in the formation of a chemical compound or alloy in the target the process is referred to as ion beam synthesis (IBS). The synthesis of SIMOX material is a two-step process, which results in the formation of a silicon on insulator (SOI) structure, as shown schematically in FIGURE 3.1. Stage 1 involves the implantation of a large dose of reactive O^+ ions into single crystal silicon to synthesise silicon dioxide. Stage 2 involves a high temperature anneal (HTA) both to annihilate lattice defects and redistribute the implanted oxygen to form a buried layer of homogeneous, stoichiometric SiO_2 with abrupt oxide-silicon interfaces. In this chapter, for clarity, only the buried stoichiometric SiO_2 layer, which evolves during the HTA, is referred to as the "buried oxide" or BOX.

By design an ideal SOI substrate is a strain and defect free, single crystal Si/amorphous SiO_2/single crystal Si (c-Si/a-SiO_2/c-Si) structure with planar, abrupt interfaces and a high quality, homogeneous BOX. With only a cursory consideration one would conclude that such structures cannot be synthesised by ion implantation due to both lattice damage caused by the momentum of the incident ions and, also, the broad Gaussian-like concentration-depth profile of the implanted oxygen. However, this would be to ignore the importance of **chemistry** and **thermodynamics** which are exploited in SIMOX technology to convert the diffuse O^+ implanted layer into a planar structure with atomically abrupt

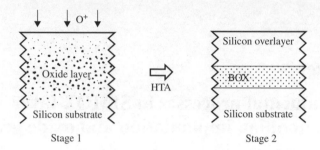

FIGURE 3.1   Schematic showing the basic SIMOX process. Stage 1: O$^+$ implantation. Stage 2: high temperature anneal (HTA). The anneal converts the broad oxide layer into a homogeneous amorphous buried oxide (BOX).

Si/SiO$_2$ interfaces whose quality approaches that of a thermal oxide/Si interface.

3.1.1   Background

Arguably the first reports that the implantation of high velocity ions into a target material could have beneficial effects were made in 1952 by Ohl [2], who irradiated point contact silicon diodes with He$^+$ ions. These experiments were followed by Coussins in 1955 [3] who bombarded germanium crystals with various ions. The first suggestion that the surface layer of a target material could be converted into an oxide or nitride by the implantation of high doses of reactive O$^+$ or N$^+$ ions was reported in 1956 by Smith [4]. Whilst the need for thermal annealing to restore the crystal structure was understood by Shockley [5], the distinction between chemical effects and damage effects was not developed until 1961, in a paper by Brodov et al. [6]. Ten years elapsed after Smith's paper before Watanabe and Tooi in 1966 [7] claimed to have synthesised SiO$_2$ by oxygen implantation and, independently, Pavlov and Shitova in 1967 [8] reported the formation of an insulating layer. Subsequently Gusev et al. [9] synthesised SiO$_2$ layers to protect p-n junctions as also reported by Fritzche and Rothemond in 1970 [10]. During the same period Freeman and colleagues [11] were investigating the chemical implications of reactive ion (O$^+$, N$^+$) implantation but concluded that the layers were of poor quality. Kelly et al. [12, 13] investigated ion implanted insulating compounds and noted that chemical bond type, temperature of crystallisation and sputtering play a role in the formation of a compound phase. They concluded that the compounds will not form if they are unstable during ion irradiation (implantation) and these workers identified SiO$_2$ as being chemically stable but structurally unstable from which they concluded that silicon dioxide would be amorphous – a feature that has been crucial in the development of SIMOX technology (in contrast to the nitride and carbide analogues which recrystallise [14]). During this same period (late 1960s to early 1970s) Schwuttke and co-workers [15, 16] reported for the first time the synthesis of deep buried insulating layers formed by O^+, N^+ or C^+ implantation at the high energy of 2 MeV. All the previous investigations had employed ions with an energy typically of a few tens keV.

During the mid 1970s several groups investigated the synthesis of SiO_2 in an endeavour to exploit IBS in order to fabricate new materials and structures for applications in microelectronics. At that time the doping of silicon by ion implantation was relatively commonplace and The Plessey Co. Ltd. was already fabricating the first all-implanted silicon bipolar devices [17]. In 1976 Dylewski and Joshi [18–20] reported experiments to investigate the electrical properties of surface oxides formed by 30 keV O^+ implantation. Similar work by Badawi and Anand [21], but using 150 keV O^+ ions, enabled these workers to synthesise a relatively good quality insulator (assessed using C-V methods) beneath a thin silicon overlayer, which prompted them to speculate that such buried oxide layers might be suitable for applications in microelectronics. Dexter et al. [22] successfully synthesised a buried nitride layer using 150 keV N^+ ions. Das and colleagues [23] (who proposed the acronym B-IMPLOX) used 200 keV O^+ ions to extend the experiments of Badawi and Anand and characterised the buried oxide using electrical methods, optical spectroscopy and cross-section transmission electron microscopy (XTEM) to optimise the processing. Gill and Wilson [24] used Rutherford backscattering to investigate the stoichiometry of the synthesised oxide. Elsewhere Astakhov et al. [25] and Kirov et al. [26] were also studying the optical and electrical properties of synthesised oxide layers. All researchers were putting great emphasis upon thermal processing of the synthesised structures as it was realised that good quality synthesised materials would only be achieved if the implantation stage is followed by a thermal anneal, as is the case when using ion implantation to dope semiconductors. However, in the case of IBS of an oxide layer beneath a single crystal silicon overlayer, it was found necessary to implant the O^+ ions into a silicon substrate maintained at an elevated temperature in order to promote dynamic annealing during implantation and thereby retain the crystallinity of the silicon overlayer.

During 1978, Izumi et al. [27] published a report of their pioneering experiments in which they synthesised a BOX using 150 keV O^+ ions to a dose of $1.2 \times 10^{18}\, O^+\, cm^{-2}$ with the substrate held at an elevated temperature during the implantation and subsequently annealed at 1150°C for two hours. The synthesised structure served as an SOI substrate upon which they subsequently

grew an epitaxial silicon layer in which they successfully fabricated a nineteen-stage CMOS ring oscillator. These workers paved
the way for a new SOI materials technology [27] and introduced the
acronym SIMOX (separation by implanted oxygen) as the name for
the technology. The current commercial success of SIMOX technology is due to these original experiments by Izumi et al. [27]
plus other key results reported during the 1980s and 1990s, which
included very high temperature annealing to achieve the complete
segregation of excess oxygen by the BOX [28], ^{18}O tracer experiments to investigate oxygen mass transport [29], identification of
the low-dose "Izumi window" [30] and recognition of the role of
an oxygen annealing ambient [31] and the temperature ramp rate
[32, 33]. These topics are discussed in Section 3.3.2 and by other
authors elsewhere in this volume.

In this section we will focus on the science and technology that
underpin the synthesis of the oxide layer and current processing
strategies used to convert the as-implanted layer into a BOX layer
with good electrical integrity and thereby form a high quality SOI
substrate suitable for advanced MOST, HBT, MEMS and photonic
applications [1].

3.1.2 The case for SOI

Near ideal performance of silicon MOS and HBT devices can only
be achieved if the devices are fabricated in a thin silicon substrate
with a thickness comparable to the depth of the device's p-n junctions, currently appreciably less than one micron for advanced
CMOS [34]. The advantages of a thin substrate include reduced
stay capacitance and junction leakage and much enhanced radiation tolerance. Whilst designers can use CAD methods to "build"
devices and circuits on thin substrates, process engineers, who
live in the real world, cannot due to the problems of handling the
fragile material. The practical solution is to provide some form
of mechanical support for the thin silicon film. Hence SOI technologies have been developed where the silicon film is supported
by either (i) a thick insulating substrate, for example silicon on
sapphire (SOS) [1], or (ii) a thin insulator such as amorphous
SiO_2 supported in turn by a bulk silicon wafer, as exemplified
by the commercially successful SIMOX, SmartCut/Unibond® and
ELTRAN® technologies [1]. These two SOI strategies are shown
schematically in FIGURE 3.2.

As a substrate for advanced CMOS circuits, which is currently
the major application of SOI, the thickness of the silicon film (t_{si})
is required to be typically 20 nm to 60 nm and the thickness of
the BOX typically 60 nm to 80 nm for fully depleted (FD-MOS)

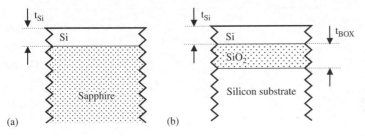

FIGURE 3.2   Schematic showing cross-sections of (a) thick insulator SOI and (b) thin insulator SOI substrates.

devices, and about 80 nm and 100 nm to 200 nm, respectively, for partially depleted (PD-MOS) devices [1].

3.2   PHENOMENOLOGY OF OXIDE SYNTHESIS BY O$^+$ ION IMPLANTATION

The number of ions (atoms) that are implanted into a target is defined by the dose (fluence) in units of ions cm^{-2} incident upon the target where the depth of penetration of the ions can be varied by adjusting their kinetic energy (keV). In the context of SIMOX technology the dose is in the range 10^{17} O$^+$ cm^{-2} to 10^{18} O$^+$ cm^{-2} and the ion energy is typically 50 keV to 200 keV.

As the incident high energy O$^+$ ions penetrate the silicon target they are randomly scattered by the lattice atoms causing the projectiles to slow down and eventually come to rest. Due to the random nature of this process the ions will not have a unique range but will come to rest with a broad distribution of ranges [35]. FIGURE 3.3 shows computed concentration-depth profiles for 65 keV and 190 keV O$^+$ ions implanted into silicon, where the target is assumed to be amorphous, to avoid anomalous effects due to channelling [36]. In the present context it is adequate to consider the distributions to be Gaussian, which can be described by a mean projected range ($\bar{R}_p$) and range straggle ($\overline{\Delta R_p}$). However, in practice the profiles are not symmetric, as is evident in FIGURE 3.3, so it is useful to define $\hat{R}_p$ as the depth of the peak of the oxygen distribution. For a precise definition of the implanted profiles it is necessary to invoke higher order moments [36].

The simulated distributions show that the mean projected ranges of 65 keV and 190 keV O$^+$ ions are significantly greater than the range straggle resulting in well defined buried layers of implanted oxygen with a volume concentration at the surface at least two orders of magnitude below the peak concentration. FIGURE 3.4 shows the dependence of $\bar{R}_p$ and $\overline{\Delta R_p}$ upon O$^+$ ion energy over

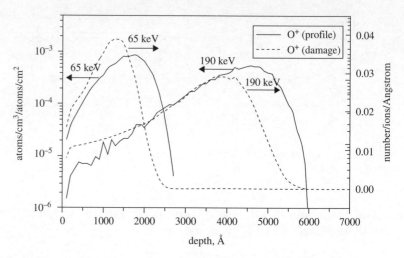

FIGURE 3.3   Simulated atomic oxygen and damage depth profiles for 65 keV and 190 keV O^+ implanted into silicon (Suspre) [35].

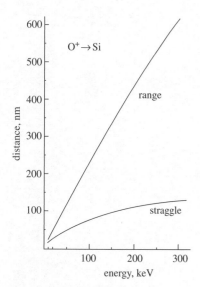

FIGURE 3.4   Projected ion range ($\bar{R}_p$) and straggle ($\overline{\Delta R_p}$) for O^+ ions implanted into silicon (Suspre) [35].

the range 10 keV to 300 keV [35]. Whilst the range increases monotonically with energy the straggle tends asymptotically to a maximum: thus the realisation of a well defined buried layer with a low oxygen concentration at the surface is more easily achieved at higher ion energies. Indeed, in the context of SIMOX technology, a shallow implanted oxygen layer demands tight control of the oxygen dose and subsequently very careful thermal processing to avoid degradation of the silicon surface.

The incident O^+ ions slow down in the silicon target by the transference of their kinetic energy to the lattice through electronic and nuclear scattering events [36]. In the present context nuclear scattering is most significant and can be modelled as hard sphere collisions of a fast O^+ ion with a stationary silicon lattice atom (FIGURE 3.5). When the O^+ ion makes a "hard" collision with a silicon atom sufficient energy may be transferred to break atomic bonds and displace the lattice atom into an interstitial site leaving behind a lattice vacancy. Thus, a Frenkel pair is formed consisting of a Si_{Int} and Si_{Vac} where the initial silicon "knock-on" atom carries forward momentum so that, on average, it will come to rest at a greater depth than the Si_{Vac} that it leaves behind. FIGURE 3.3 includes simulated vacancy distributions (damage profiles) due to the 65 keV and 190 keV O^+ implantation into silicon. Typically the depth of the damage peak ($\hat{R}_D$) is located at about 80% of $\hat{R}_p$.

When an incident O^+ ion makes a near "head-on" collision with a lattice atom sufficient momentum may be transferred to the "knock-on" silicon atom (silicon recoil) to cause second, third and higher order displacements to occur and create a small disordered volume or collision cascade (FIGURE 3.6). Such cascades can range from a volume of highly defective silicon to a small amorphous region [38]. With increasing dose the number and concentration of the cascades will increase and adjacent cascades will overlap. As this occurs the overlapping amorphous regions will start to coalesce until it becomes energetically favourable for the implanted layer to transform into an amorphous phase. This amorphisation will initially occur at the peak of the damage profile (shown in FIGURE 3.3), and then broaden until the complete surface layer is amorphised. Phenomenologically, it is appropriate to consider a "deposited energy density threshold", which is proportional to the ion dose, above which the lattice will be amorphised; this concept has been discussed by Jones et al. [39]. The value of this threshold energy (dose) is highly temperature sensitive, increasing with increasing temperature.

In the context of SIMOX technology it is important to inhibit amorphisation. Vook et al. [40] have estimated that for a dose of 10^{18} O^+ cm^{-2} at 200 keV an implantation temperature in the range 400°C to 500°C is required to retain the crystallinity of the silicon surface layer, which has been confirmed by experiment. FIGURE 3.7 shows the temperature dependence of the O^+ dose to

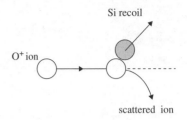

FIGURE 3.5 Schematic of a nuclear collision of an incident energetic (O^+) ion and a lattice atom (silicon).

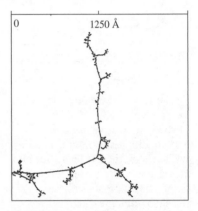

FIGURE 3.6 Monte-Carlo (TRIM) simulation of the trajectory of a 200 keV O^+ ion in silicon showing the random creation of collision cascades [37].

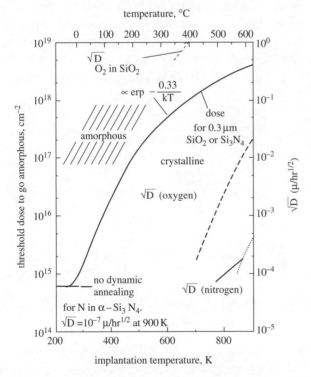

FIGURE 3.7 Temperature dependence of the threshold dose to form a homogeneous amorphous layer in silicon. Diffusivities for oxygen and nitrogen are included [40].

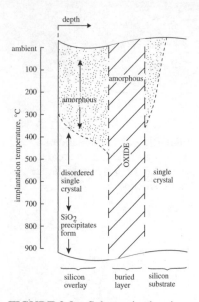

FIGURE 3.8   Schematic showing the depth dependence of the SIMOX microstructure upon implantation temperature. The free surface is on the left-hand side and increasing depth is from left to right. The amorphous phase shrinks with increasing implantation temperature.

amorphise silicon [40]. As an example, FIGURE 3.8 is a schematic showing the extent of the amorphous phase as a function of implantation temperature for a dose of 1.8×10^{18} O^+ cm^{-2} at 200 keV. Essentially the thickness of the "as-implanted" thick oxide layer ($\approx$3500 Å) is independent of the implantation temperature. However the microstructure of the silicon on either side of the oxide layer is highly dependent upon temperature, as shown. At ambient temperature ($\approx$20°C) the whole structure is amorphous to a depth of approximately ($\bar{R}_p + 2\overline{\Delta R_p}$) but with increasing temperature and the onset of dynamic annealing the amorphous layers above and below the oxide shrink until, at about 500°C, all of the implanted silicon retains its single crystal nature although it contains many crystallographic defects and a high concentration of small oxide precipitates. With increasing temperature the precipitates grow, as discussed in Section 3.3.1, and eventually become the dominant feature of the silicon overlayer. Experiments have shown that 500°C to 600°C is the optimum range of implantation temperatures for the commercial production of low defect density SIMOX substrates.

The accumulation of damage during O^+ implantation has been investigated by Holland et al. [41] who found a depth dependent volumetric change due to the high concentration of included oxygen atoms and the spatial separation of the Si_{Int} and Si_{Vac}, described above. FIGURE 3.9 shows the simulated volume change as a function of depth superimposed upon the oxygen distribution of 1×10^{18} O^+ cm^{-2} at 450 keV. They conclude that the total Si_{Int}

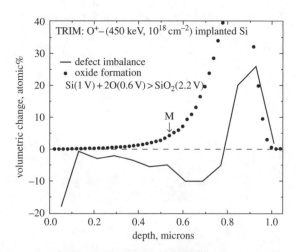

FIGURE 3.9   Monte-Carlo (TRIM) simulation showing the volumetric change in a SIMOX SOI structure arising from point defect (Si_{Int}, Si_{Vac}) separation. The arrow (M) indicates the depth of the null point between density increasing and decreasing [41].

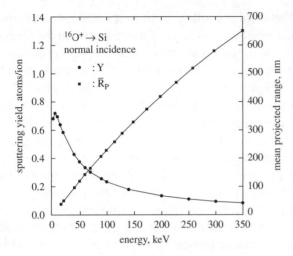

FIGURE 3.10   The energy dependence of the sputter yield of $^{16}O^+$ implanted into silicon. These data have been generated by the PROFILE Code and PRAL-87 [42].

and Si_{Vac} profiles are spatially displaced with a vacancy excess of several atomic percent dominant to a depth of $0.8\,\mu m$ and a narrow distribution of excess Si_{Int} at greater depths. These excess Si_{Int} are located in the lower wing of the oxygen distribution. Consideration of the net effect of the implanted oxygen and excess vacancies shows a balance at a depth of about $0.55\,\mu m$ (labelled M). These simulations show that during IBS the silicon overlayer above the synthesised oxide is vacancy rich whilst just below the oxide there is an excess of Si_{Int}.

A further consequence of the fast ion-atom ($O^+ \rightarrow Si$) collisional events is sputter erosion of the target which occurs when backscattered particles originating in the collision cascades are in the vicinity of the surface and have sufficient energy to overcome the surface binding energy [36]. Sputtering causes the surface to recede at a characteristic rate defined by the sputter yield (atoms lost per incident ion) through the loss of host lattice atoms (Si) and to a lesser extent the implanted species (oxygen). Experiments show the sputtering yield to be a function of ion energy and mass and crystal orientation [38]. Measured sputter yields for $200\,keV$ O^+ into silicon and thermal SiO_2 are 0.18 atoms per O^+ ion and ≈ 1.1 atoms per O^+ ion, respectively (FIGURE 3.10) [42].

FIGURE 3.11 is a schematic showing depth profiles of $200\,keV$ O^+ implanted into silicon over the dose range 1×10^{11} O^+ cm^{-2} to 2×10^{18} O^+ cm^{-2}. The solid solubility limit (SSL) at the implantation temperature and the oxygen volume concentration in stoichiometric SiO_2 (4.4×10^{22} $O\,cm^{-3}$) are indicated by the dotted lines. For the lowest dose the oxygen volume concentration

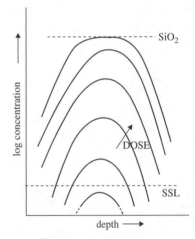

FIGURE 3.11   Schematic showing the evolution of the oxygen concentration–depth profile with increasing dose of implanted O^+ in silicon. The dotted lines correspond to the solid solubility limit (SSL) and oxygen concentration in stoichiometric SiO_2.

is below the solid solubility limit and the implanted oxygen may be considered to be a dissolved trace impurity in the silicon matrix. As the dose is increased an increasing proportion of the implanted oxygen exceeds the solid solubility limit and nucleation of second phase (oxide) precipitates will occur at these depths. The precipitates will vary in size where the largest are located at the peak of the concentration profile. For doses greater than about 10^{16} O^+ cm^{-2} significant sputtering will also occur and for still higher doses the included oxide precipitates will grow and cause significant swelling. These processes are the cause of the broadening of the oxygen profile. Still more pronounced broadening at the peak of the profile occurs for the highest dose when the maximum volume concentration of oxygen saturates at the value in stoichiometric SiO_2. This behaviour is a characteristic of the O^+/Si system and is a result of the O-Si bond strength and high diffusivity of excess oxygen in amorphous SiO_2 (FIGURE 3.7). Experiments have shown that once a stoichiometric oxide has been formed excess oxygen, due to further incremental doses, rapidly redistributes to the wings of the distribution where there are unsatisfied silicon bonds. Marker [43] and O^{18} tracer [44] experiments have shown that during implantation the excess oxygen preferentially diffuses to the upper Si/SiO$_2$ interface. Redistribution of the implanted excess oxygen is found to occur even during O^+ implantation at liquid nitrogen temperatures [45] although at liquid helium temperatures there is evidence that the excess oxygen can be accommodated in the oxide matrix but rapidly redistributes during warm-up to room temperature [46]. All experiments have shown that the synthesised oxide layer consists of amorphous SiO_2. This is a feature that is not found in the analogous N^+/Si and C^+/Si systems where the concentration of the solute atoms can exceed the stoichiometric value and the synthesised silicon nitride and carbide materials are found to be polycrystalline [14].

FIGURES 3.12(a) and (b) are schematics showing the evolution of the oxide layer with increasing dose of 200 keV O^+ ions into a silicon substrate at a nominal temperature of 500°C. These models draw upon experimental data reported in the literature and incorporate the processes of swelling and sputtering. In both figures the O^+ dose increases from left to right but with a highly non-linear scale. The symbols $\hat{R}_{OX}$ and $\hat{R}_D$ represent the depths of the peaks of the oxygen and damage profiles, respectively, whilst ϕ_c is the critical O^+ dose that just achieves the volume concentration of stoichiometric SiO_2 at the peak of the atomic oxygen profile. As the implantation temperature is 500°C the silicon above and below the synthesised buried layer of SiO_2 retains its crystallinity although it is highly defective (see FIGURES 3.7 and 3.8). The spatial distribution and density of defects is represented by the black dots.

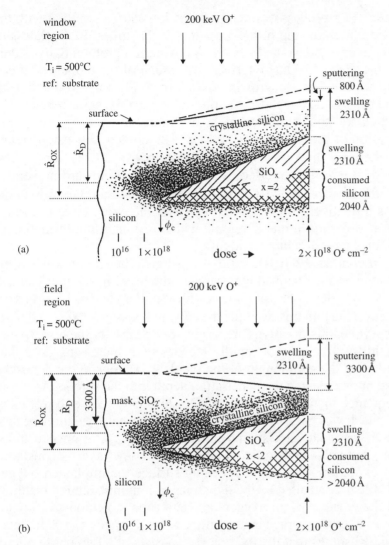

FIGURE 3.12   Schematics showing the evolution of the buried oxide with increasing dose of 200 keV O^+ ions (a) when implanted into bulk silicon with a sputter yield of 0.18 atoms ion^{-1} and (b) when implanted through a thin SiO_2 capping layer with a sputter yield of 1.1 atoms ion^{-1}. The dose scale is highly non-linear and the indicated thicknesses relate to a final dose of 2×10^{18} O^+ cm^{-2}.

In FIGURE 3.12(a) the O^+ ions are implanted into bare silicon with an experimentally determined sputter yield of 0.18 atoms per O^+ ion [42]. For doses below ϕ_c the structure is dominated by lattice defects and small oxide precipitates, as discussed in Section 3.3.1. For doses above about 1×10^{16} O^+ cm^{-2} sputter erosion of the surface becomes progressively more significant. At a dose of ϕ_c a continuous stoichiometric buried oxide is just achieved which increases in thickness with further increases of

29

dose. This process of internal oxidation causes swelling with the oxide layer having a thickness that is 2.2 times the thickness of the layer of consumed silicon. As oxidation preferentially occurs at the upper Si/SiO_2 interface [43] much of the damaged silicon is consumed by the growing oxide. In contrast the lower SiO_2/Si interface is unchanged and lattice defects in this region remain. The schematic shows separately the extent of the swelling and sputter erosion of the free silicon surface. The quoted dimensions relate to the layer thicknesses for a total dose of 2×10^{18} O^+ cm^{-2}. It is evident that the net effect is swelling with an upward movement of the surface of $1510\,Å$. In this case, where the target material has a relatively small sputtering yield, the oxide layer is formed in a "diffusion limited" regime which results in the formation of amorphous, stoichiometric SiO_2.

In contrast FIGURE 3.12(b) illustrates the situation when the silicon wafer is capped with a material with a high sputter yield, in this case a deposited SiO_2 mask with a sputter yield of 1.1 atoms per O^+ ion [42]. In this example the cap thickness is $3000\,Å$. All other experimental conditions are as above. Because of the high sputter yield the free surface rapidly recedes with increasing dose thus reducing the path length of the incident O^+ ions in the deposited cap but increasing the penetration depth into the silicon. As a consequence incremental doses of O^+, above ϕ_c, come to rest close to the lower SiO_2/Si interface where internal oxidation occurs. In this case damaged silicon above the oxide layer is not consumed but, instead, the disordered silicon below the layer is consumed by oxidation. Due to the receding surface the implanted oxygen profile is controlled by the dynamics of ion implantation resulting in the creation of an oxygen deficient (non-stoichiometric) oxide layer with an average composition of SiO_x where $x < 2$. In detail XTEM shows that the oxide layer consists of a two phase microstructure of stoichiometric SiO_2 precipitates in a supersaturated silicon matrix. This is conveyed in the schematic by a greater thickness of silicon consumed ($>2040\,Å$) than in the diffusion limited case (FIGURE 3.12(a)). Upon achieving the final dose of 2×10^{18} O^+ cm^{-2} the free surface has receded by $990\,Å$ and the synthesis of the non-stoichiometric oxide has occured in a "sputter limited" regime.

3.3 PHYSICS AND CHEMISTRY OF SIMOX LAYER FORMATION

The build up of the oxygen profile has been described above; however, the synthesis of the layer of oxide during ion implantation is mediated by the establishment of Si-O bonds, the presence of excess point defects (Si_{Int} and Si_{Vac}) created in the collisional cascades (FIGURE 3.6) and the generation of an atomic volume to accommodate the oxide. The atomic volume of an SiO_2 molecule is about 2.2 times the volume of the single constituent silicon atom [47].

3.3.1 Step 1: ion implantation

Synthesis of an oxide layer during implantation depends upon four overlapping stages [48], namely:

(i) supersaturation, due to the build up of the oxygen concentration with increasing O^+ dose,

(ii) nucleation of oxide precipitates (chemical bonding) during implantation,

(iii) oxide precipitate growth (Ostwald ripening),

(iv) coalescence of the oxide precipitates.

These stages are discussed below.

Supersaturation

The temperature dependence of the solubility of oxygen in silicon is shown in FIGURE 3.13, due to Borghesi et al. [49]. The solubility can be expressed as:

$$C(T) = C_o \exp\left(-\frac{DE}{kT}\right)$$

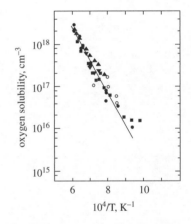

FIGURE 3.13 Oxygen solubility in silicon as a function of temperature [49].

where $C_o = 9 \times 10^{22}$ cm^{-3} and $DE = 1.52$ eV.

The solubility at a typical implantation temperature of 600°C is $\approx 10^{17}$ O cm^{-3}. In the present context the concentration at the peak of the depth profile exceeds the solid solubility limit for doses greater than 1×10^{12} O^+ cm^{-2} to 5×10^{12} O^+ cm^{-2} over the energy range 20 keV to 200 keV.

The volume concentration (N) of the implanted oxygen is approximately [38]:

$$N = \frac{0.4N_{sq}}{\overline{\Delta R_p}}$$

where N_{sq} is the implanted dose (O^+ cm^{-2}) and $\overline{\Delta R_p}$ is the range straggle. Using this simplistic formula, assuming no sputtering or diffusional broadening, it is evident that a dose greater than 5×10^{12} O^+ cm^{-2} of 200 keV O^+ ions will give a volume concentration in excess of the solid solubility limit. However, direct synthesis of an oxide layer during ion implantation requires the peak of the oxygen profile to achieve the volume concentration of oxygen in the stoichiometric oxide ($\sim 4.4 \times 10^{22}$ cm^{-3}), which requires a dose of approximately 1×10^{18} O$^+$ cm^{-2}, as shown schematically in FIGURE 3.11.

Precipitate nucleation during implantation

The dominant mechanism for the nucleation of SiO$_2$ precipitates under O$^+$ implantation is found to be dependent upon the processing condition and physical state of the surface of the silicon wafer [50]. For low O$^+$ doses and a high implantation temperature heterogeneous precipitation preferentially occurs at nucleation sites that Bourret [51] has suggested are vacancy clusters located in the vicinity of the implantation damage peak. Evidence is forthcoming from low dose fully annealed structures as XTEM shows a band of precipitates at depth R$_D$ with a second more dominant band of precipitates close to the peak of the oxygen profile (see Section 3.3.2).

The nucleation of second phase precipitates has been discussed by Mantl [48] who points out that nucleation occurs under extreme non-equilibrium conditions in a silicon rich environment (oxygen in a silicon matrix). The implanted oxygen atoms are required to diffuse a short distance to sites where atomic bonding, nucleation and subsequent growth of the precipitate will occur.

Experiments indicate that the evolution of the precipitates is controlled by irradiation dependent physical processes, including [48]:

(a) radiation moderated diffusion,
(b) dissolution of precipitates due to knock-on solute atoms (oxygen),
(c) dissolution due to mixing in the collision cascade,
(d) solute enrichment due to the point defect flux,
(e) precipitate nucleation at radiation induced lattice defects.

Reference to the literature shows that during ion implantation the density and size of the synthesised precipitates are highly dependent upon implantation temperature, dose, ion flux and beam scan frequency [52]. Indeed lack of control of these parameters was responsible for the variable quality of SIMOX materials reported in the literature prior to the advent of dedicated O$^+$ ion implanters.

The thermodynamics of the nucleation and growth of precipitates in a supersaturated system formed by ion implantation is driven by the free energy of the system, as discussed by Mantl [48]. In the absence of ion implantation effects the change in Gibbs free energy (ΔG) when a new phase of volume V and interface area A is nucleated is [53]:

$$\Delta G = V(\Delta G_V + \Delta G_E) + A\gamma$$

where ΔG is the change in free energy per unit volume occupied by the new phase, ΔG_E is the increase in elastic strain energy per unit volume of the precipitate and γ is the specific interface energy. The difference in the volumes of the unstressed matrix and unstressed precipitate (ΔV) is the cause of internal strain where ΔG_E depends upon ΔV and the elastic constancy of the matrix and precipitate and, also, the precipitate morphology. FIGURE 3.14 shows the dependence of ΔG_V on the radius of spherical precipitates at two difference temperatures where r_c is the critical radius which may be calculated using [48]:

$$r_c = \frac{2\gamma}{\Delta G_V + \Delta G_E}$$

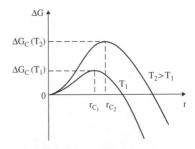

FIGURE 3.14 Schematic showing the change in Gibbs free energy (ΔG) for a spherical precipitate in a supersaturated matrix as a function of precipitate radius at temperatures T_1 and T_2 where $T_2 > T_1$ [48].

When, at a given temperature, the system moves to reduce its energy those precipitates of a radius less than r_c are unstable and will tend to dissolve. From the figure the critical radius above which the precipitates are stable increases with increasing temperature.

The chemistry of oxide synthesis has been discussed by Jaussaud et al. [52] in an investigation of the extended defects found in SIMOX materials. These workers considered the collision cascades occurring during O^+ implantation as a contributory source of point defects (Frenkel pairs):

$$Si \rightarrow Si_{Int} + Si_{Vac} \tag{3.1}$$

Further they invoke the emission of silicon interstitials, which is a consequence of the oxidation of silicon [52], as the vehicle for the formation of the "accommodation volume" that is required if the larger SiO_2 molecules are to be accommodated without disruption of the matrix:

$$(1 + x)Si + 2O_{Int} \rightarrow xSi_{Int} + SiO_2 \tag{3.2}$$

and by combining EQNS (3.1) and (3.2) they describe the IBS of oxide precipitates by:

$$(1 + x)(Si_{Int} + Si_{Vac}) + 2O_{Int} \rightarrow xSi_{Int} + SiO_2$$

where $x = 1.25$. In common with the thermal oxidation of bulk silicon there is an emission of $0.63Si_{Int}$ per (implanted) oxygen atom. As the collisional cascades maintain an excess population of vacancies, oxide growth by IBS is easier than thermal oxidation [54]. As shown in FIGURE 3.9 spatial separation of the Si_{Int} and Si_{Vac} populations occurs and, thus, the reaction rate will have a depth dependence, which is confirmed by experiment and illustrated schematically in FIGURE 3.12(a). Kilner et al. [44] have shown by O^{18} tracer experiments that the existence of a continuous buried oxide blocks the passage of silicon atoms to the substrate which, therefore, cannot act as an efficient sink for Si_{Int}. Oxidation occurs preferentially at the upper Si/SiO_2 interface as illustrated schematically in FIGURE 3.12(a).

The synthesis of each SiO_2 molecule causes the ejection of $0.63Si_{Int}$ which must be eliminated if distortion of the matrix is to be avoided. Stoemenos [55] has shown that this can be efficiently achieved if the Si_{Int} are able to migrate to the free surface where they are incorporated by "internal epitaxy". These workers note that incorporation at the surface will be dependent upon the chemical state of the surface especially the presence of contaminants and a capping layer. The effect of the capping layer during high temperature annealing is discussed in Section 3.3.2.

Growth of oxide precipitates

Growth of the oxide precipitates during both implantation and subsequent thermal processing is controlled by the Si/SiO_2 interfacial energy (γ) and local transport of oxygen and point defects. Experiments show that the competitive growth of precipitates is controlled by diffusion with the larger precipitates growing at the expense of the smaller ones (diffusion controlled precipitate growth – Ostwald ripening [56]), the driving force being a reduction of the total free energy.

Mantl has shown that the difference in the chemical potential of precipitates of different size gives rise to a dependence of the solute (oxygen) equilibrium concentration around those precipitates. The concentration ($c_s(r)$) at the surface of a precipitate of radius r, given by the Gibbs-Thomson equation [57], is shown schematically in FIGURE 3.15 for precipitates of radius r_1 and r_2 where $r_1 > r_c > r_2$. The concentration remote from the precipitates is denoted by c_∞ and thus there is local supersaturation around each precipitate. The concentration gradient between the two precipitates drives the oxygen diffusion to cause dissolution of the small precipitate with growth of the larger one. In general during IBS the critical radius is proportional to the wafer temperature and inversely proportional to the degree of oxygen supersaturation.

FIGURE 3.16, due to Borghesi et al. [58], shows the temperature dependence of the critical radius of oxide precipitates in bulk

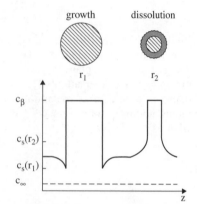

FIGURE 3.15 Schematic showing the solute (oxygen) concentration in the vicinity of a growing ($r_1 > r_c$) and shrinking ($r_2 < r_c$) precipitate [48].

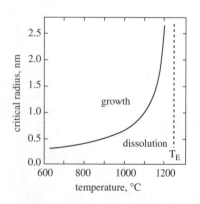

FIGURE 3.16 Temperature dependence of the critical radius (r_c) of oxide precipitates in silicon as a function of temperature [58].

silicon. The value increases rapidly with increasing temperature, where T_E is the temperature at which the system is in equilibrium. At any given temperature precipitates with a radius lying above the curve grow at the expense of those with a radius below the curve.

As pointed out by Reeson et al. [54], a detailed analysis of IBS is complex as there are additional factors to consider due to the presence of radiation defects and the increase in oxygen concentration with each increment of dose of O^+ ions.

Coalescence of oxide precipitates

When the density of precipitates is high they may be in close proximity to each other or even touch, as shown in FIGURE 3.17. In this situation coalescence of adjacent precipitates may occur as the system moves to lower its free energy [59]. The functional dependence of the oxygen concentration at the neck of the two touching spheres is similar to the Gibbs-Thomson equation [53] and due to the negative curvature at the neck (ρ), the concentration is smaller at the neck than in the surrounding matrix. Consequently the oxygen flux (J) is towards the neck and is approximated by [59]:

$$J = \frac{\gamma c_\infty D V_a}{\rho^2 kT}$$

where D is the oxygen diffusion coefficient, ρ the radius of the neck and V_a the atomic volume. Coalescence is technologically important as it is the mechanism which facilitates the lateral growth of precipitates and the eventual transformation, during the high temperature anneal, of the two phase structure into a continuous oxide layer – the BOX, which is the key to the success of SIMOX technology. Once formed the BOX can be considered to be a single thermally stable precipitate of infinite radius.

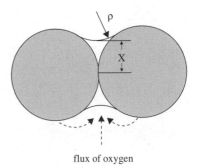

flux of oxygen

FIGURE 3.17   Schematic showing the solute flux (oxygen) in the vicinity of two coalescing precipitates (SiO_2) in a silicon matrix [48].

3.3.2   Step 2: high temperature anneal

The purpose of the high temperature anneal (HTA) [60] is to thermally activate chemical processes (i) to annihilate lattice defects and (ii) to convert the diffuse oxygen distribution produced during Stage 1 into a planar SOI structure with abrupt Si/SiO_2 interfaces that define the thickness of the BOX, shown schematically in FIGURE 3.1. This is achieved by mass transport of oxygen from the wings of the implanted oxygen distribution (regions I and III in FIGURE 3.18(a)(i)) to the central region by the physical processes of diffusion controlled precipitate growth (Ostwald ripening) [56] and precipitate coalescence, already discussed in Section 3.3.1. During Stage 2 the mass transport is thermally driven but mediated by the excess point defects that are a memory of the ion

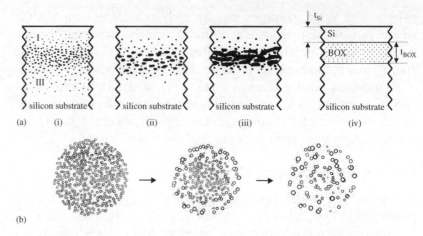

FIGURE 3.18 Schematic showing the growth (Ostwald ripening) and coalescence of SiO_2 precipitates with increasing anneal temperature. These processes cause the broad oxygen implanted layer to transform into a BOX with abrupt Si/SiO_2 interfaces. (a) The regions labelled I and III correspond to the wings of the SIMS oxygen profiles from as-implanted samples (see Figures 3.19 and 3.22). (b) Computer simulations of Ostwald ripening showing the evolution of an ensemble of spatially inhomogeneous precipitates after various time intervals [61].

implantation stage. FIGURE 3.9 shows typical distributions of the excess point defects where Si_{Vac} and Si_{Int} are spatially separated.

During the original SIMOX experiments in the 1970s and 1980s the HTA was limited to a temperature in the range 1150°C to 1200°C as this was the upper limit of the available clean room furnaces. These temperatures were only sufficient to dissolve the smallest oxide precipitates (see FIGURE 3.16) and achieve partial redistribution of the implanted oxygen. It was not until 1985 that the first reports of the complete segregation of the implanted oxygen on to the BOX layer entered the literature. The relevant experiments included 1300–1350°C anneals of small SIMOX samples in a sealed ampoule at LETI [28], 1200°C anneals of analogous N^+/Si SIMNI samples at the University of Surrey [62] and 1405°C lamp anneals of SIMOX wafers at Bell Labs [63]. In each case planar SOI structures with abrupt interfaces were reported. The absence of clean room furnaces and furnace tubes able to operate at and above 1350°C, which were free of metal contaminants, initially constrained commercially orientated research in very high temperature annealing. High temperature annealing facilities are now available; however, although the need for a very high thermal budget is common knowledge, the details of the ramp rate, time-temperature profile, anneal ambient and the nature of a cap, if used, all tend to be laboratory specific and commercially sensitive. Published temperature and time parameters necessary to achieve precipitate dissolution in the silicon overlayer are typically 1350°C for 6 hr in a weakly oxidising ambient of Ar + 0.5% O_2, to minimise thermal etching of the free silicon surface [60, 64].

Annealing ambient and surface capping layer

During the HTA it is important to avoid unintentional thermal oxidation of the free silicon surface as this will cause thinning of the silicon overlayer. Also, it is necessary to inhibit the formation of silicon monoxide in order to avoid thermal etching and roughening of the surface [64]. Control of the surface chemistry is normally achieved by annealing in a weakly oxidising ambient. An alternative method adopted to protect the free surface and to inhibit unintentional contamination is to cap the surface, for example with a layer of deposited TEOS silicon dioxide. Jiao et al. [65] have investigated the effects of a capping layer in terms of its control of internal thermal oxidation. They report that the presence of a cap inhibits mass transport of oxygen from the ambient (Ar + O_2) to the evolving BOX layer, which consequently is thinner than if the sample was annealed without a cap. Thus the absence of a cap permits internal oxidation to occur with thickening of the BOX layer, which is discussed below in terms of the ITOX process [66]. Typically, they find it is necessary to implant a marginally larger O^+ dose if a thin film SIMOX structure is to be annealed with a deposited cap. Jiao et al. [65] report a further disadvantage of a deposited cap, namely that it diminishes the effectiveness of the free silicon surface to act as a sink for excess Si_{Int} which results in a higher defect density in the silicon overlayer.

Nakashima [67] has also investigated the effects of an oxidising ambient upon the growth and dissolution of oxide precipitates located in the silicon overlayer. He found that the density and mean size of precipitates could be varied in a predictable manner by adjusting the oxygen content of the ambient gas (Ar + x% O_2), annealing time and temperature. He reported that for high oxygen concentrations precipitates grow in size whilst for low oxygen concentrations precipitate dissolution occurs [34]. These observations paved the way for the internal oxidation (ITOX) process [66], discussed below.

Johnson et al. [64] have reported that (i) the presence of a deposited silicon dioxide cap and also (ii) a pre-existing thick buried oxide layer will influence the redistribution of oxygen during the HTA as each will act as a precipitate of infinite radius and getter oxygen from the silicon overlayer (region I in FIGURE 3.18). The authors find that the annealing ambient, the temperature ramp rate and the presence or absence of a cap all influence the formation of the BOX layer and the density of extended defects in the overlayer.

Tan et al. [68] find that the redistribution of oxygen can be controlled by varying the thermal ramp rate and report that a slow ramp rate gives the smallest oxide precipitates sufficient time to grow to a stable size prior to the next temperature increment but, undesirably, allows Si_{Int} to aggregate into silicon islands inside the BOX layer. In contrast a fast ramp rate inhibits the mass transport of Si_{Int} and prevents the growth of silicon islands, provided that the implanted dose is close to the "Izumi window" [30, 69], discussed below. Jeoung et al. [70] report that for low energy, thin SIMOX structures the initial growth of the BOX occurs at a lower anneal temperature due to the smaller straggle and hence shorter distances for the mass transport of oxygen.

Thick BOX ($\phi \gg \phi_c$)

The original SIMOX experiments involved the implantation of "high" doses of about 2×10^{18} O^+ cm^{-2} in the energy range 150 keV to 200 keV which directly synthesises a thick oxide layer. During the HTA a good quality SOI structure is formed with typical layer thicknesses of $t_{BOX} \approx 4000$ Å and $t_{Si} \approx 3000$ Å. For these samples the HTA is relatively straightforward as a continuous oxide layer exists prior to the anneal; however, the anneal conditions do impact upon the evolution of extended defects in the overlayer, a topic discussed by other authors in this volume. Because of the long implantation times required to achieve such high doses, hence high wafer cost, these materials tend not to be used as SOI substrates for commercial production of advanced CMOS/SOI circuits; however, they still have currency for radiation hard circuits [1].

FIGURE 3.19 shows SIMS depth profiles from samples implanted with a dose of 1×10^{18} O^+ cm^{-2} at an energy of 70 keV where curve (a) was recorded from an as-implanted sample and curve (b) after HTA [71]. The as-implanted profile confirms that a buried layer of stoichiometric SiO_2 was formed during the implantation (see FIGURE 3.12(a)) and that the oxygen has a broad distribution, as indicated by the wings of the SIMS profile (labelled I and III). During the HTA a dramatic redistribution of oxygen occurs, with oxygen diffusing from the wings to the oxide layer which consequently grows in width and develops abrupt Si/SiO_2 interfaces. The oxygen signal in the vicinity of the free surface is mainly due to the presence of a thermal oxide; independent measurements show that essentially all of the implanted oxygen has been gettered by the BOX layer.

FIGURE 3.20 shows typical XTEM micrographs from a high dose sample implanted at a temperature $>500°C$ with 1.8×10^{18} O^+ cm^{-2} at 200 keV (a) before and (b) after HTA [72].

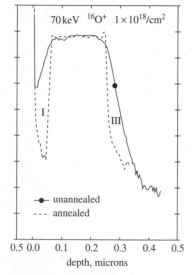

70 keV $^{16}O^+$ $1 \times 10^{18}/cm^2$

I

III

—•— unannealed
- - - annealed

0.5 0.0   0.1   0.2   0.3   0.4   0.5
depth, microns

FIGURE 3.19   SIMS oxygen profiles from samples implanted with 1×10^{18} O^+ cm^{-2} at 70 keV before and after an HTA at 1405°C for thirty minutes. The wings of the oxygen distribution are labelled I and III [71].

As the dose is much greater than ϕ_c a continuous homogeneous oxide layer has been directly synthesised beneath a highly defective single crystal silicon overlay. Before the HTA the Si/oxide interfaces, defining the oxide layer, are irregular where the lower interface shows greater irregularity due to a striated $SiO_2/Si/SiO_2$ microstructure. The silicon overlayer contains a high concentration of small oxide precipitates. During the HTA ($>1300°C$, 4 hr, Ar + 0.5% O_2) dissolution of the smaller precipitates occurs releasing oxygen which mainly diffuses to the buried oxide causing the thickness to increase. Under these typical processing conditions the measured layer thicknesses are $t_{Box} \sim 4000\,\text{Å}$ and $t_{Si} \sim 3000\,\text{Å}$ [72]. A characteristic of the resulting thick BOX layer is the presence of faceted silicon islands adjacent to the lower BOX/Si interface, which can be seen in the figure and are a consequence both of the excess Si_{Int} created during the implantation (see FIGURE 3.9) and of the thermal evolution of the striated $SiO_2/Si/SiO_2$ microstructure seen in the as-implanted micrograph.

Jiao et al. [65] have carried out a detailed study of SIMOX materials and their results are used to further illustrate the evolution of a thick BOX. FIGURE 3.21 shows samples implanted with $7 \times 10^{17}\,O^+\,\text{cm}^{-2}$ at 65 keV before and after annealing. Due to the smaller straggle (ΔR_p) at this energy the dose is still greater than ϕ_c and a continuous oxide is directly synthesised during Stage 1. In this sample the striations at the lower Si/oxide interface are clearly visible, indicated by arrows. After the HTA a well defined homogeneous BOX is formed but again with silicon islands trapped in the oxide in the vicinity of the lower interface. The upper interface is free of trapped silicon islands as the incremental doses of O^+ during the implantation are deposited in the vicinity of that interface (see FIGURE 3.12(a)) and oxygen diffuses to satisfy any free silicon bonds to give a relatively abrupt interface even before the HTA [43]. Jiao et al. [65] have further investigated the dependence of the SIMOX structures upon lower O^+ doses and find that after an HTA the density, location and size of the silicon islands trapped in the BOX are very sensitive to the dose. A broad generalisation is that they found the density of silicon islands to be inversely related to the dose. Thus, under the processing conditions discussed here, it is necessary to implant O^+ doses significantly greater than ϕ_c to achieve a thick BOX layer that has good electrical integrity.

Thin BOX ($\phi < \phi_c$)

Thin BOX layers with good electrical integrity cannot simply be formed by reducing the implanted O^+ dose to a value just greater than ϕ_c due to the formation of a high density of included silicon

(a) as-implanted

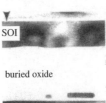

(b) annealed

FIGURE 3.20 Typical cross-section TEM micrographs of high dose SIMOX samples (a) as-implanted and (b) after an HTA [72].

FIGURE 3.21 Cross-sectional TEM micrographs of SIMOX samples implanted at 65 keV with a dose of $7.0 \times 10^{17}\,O^+\,\text{cm}^{-2}$ (c) before and (c-a) after HTA at 1350°C for 4 hr in Ar(O_2) without a cap. Figure reproduced from [65].

islands during the HTA [65]. This is bad news for a strategy of reducing the cost of SIMOX substrates simply by reducing the dose. However, there is a solution, first discussed by Izumi et al. during 1993 [30], as experiments show that there is an energy dependent low-dose window ("Izumi window") below ϕ_c for which a thin continuous BOX may be formed during the HTA. This provides a viable route to produce lower cost, thin film SOI substrates with good electrical viability. The challenge for the SIMOX/SOI vendors is to optimise and stabilise processing conditions in order to exploit this rather narrow low-dose "Izumi window" [73].

It is shown in Section 3.2 that implanted O^+ doses less than ϕ_c have a net Gaussian-like concentration depth distribution with the oxygen mainly contained in second phase (SiO_2) oxide precipitates, the largest precipitates being located at the peak of the oxygen concentration profile. During the HTA the process of Ostwald ripening [56] (see Section 3.3.1) causes the small precipitates in the wings of the concentration profile to dissolve and the large precipitates to grow. The net effect is mass transport of oxygen to the peak of the atomic oxygen distribution where the precipitates will eventually coalesce. This process is, in part, mediated by the presence of lattice defects which can themselves getter the diffusing oxygen. As the peak of the damage distribution lies closer to the surface (see FIGURE 3.3) the lattice defects therein can facilitate precipitate nucleation and give rise to a second discrete band of coalescing precipitates [33]. The consequences of Ostwald ripening are illustrated in FIGURE 3.22 which shows SIMS depth profiles from samples implanted with an O^+ dose below ϕ_c, namely 3.3×10^{17} O^+ cm^{-2} at 70 keV where curve (a) is from an "as-implanted" sample and curve (b) after HTA (1320°C, 6 hr) [74]. It is evident that under the particular processing conditions the anneal has caused the oxygen in the wings of the implanted profile (labelled I and III) to segregate to the peak of the atomic oxygen distribution at a depth $\hat{R}_p$ where the largest precipitates coalesce and form a BOX layer. Prior to the anneal a continuous homogeneous oxide layer did not exist. Independent measurements of these and similar samples after the HTA confirm the BOX to be amorphous stoichiometric SiO_2.

FIGURES 3.23(a) and (b) from Li et al. [29] show typical XTEM micrographs from samples implanted at a temperature of ~680°C with 4.3×10^{17} O^+ cm^{-2} at 90 keV before and after an HTA. The as-implanted sample is still a single crystal but with a broad band of damage with lattice defects lying on {311} and {111} planes and includes a distribution of oxide precipitates, which give rise to the upper and lower wings of the SIMS distribution shown in FIGURE 3.22. FIGURE 3.23(b) shows the structure after an HTA (1360°C for 6 hr) which causes the oxygen to redistribute

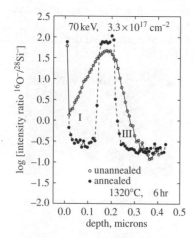

FIGURE 3.22 SIMS oxygen profiles from samples implanted with 3.3×10^{17} O^+ cm^{-2} at 70 keV (a) before annealing and (b) after HTA at 1320°C for 6 hr [74]. The wings of the oxygen distribution are labelled I and III [74].

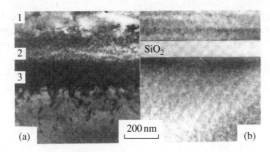

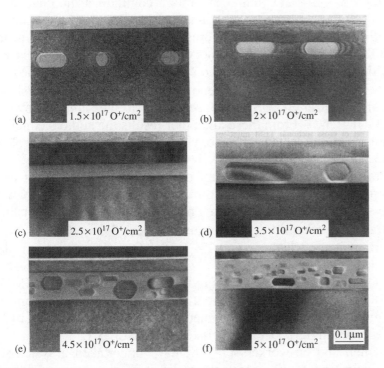

FIGURE 3.23   Cross-sectional TEM micrographs of SIMOX samples implanted with 4.3×10^{17} O^+ cm^{-2} at 90 keV (a) before and (b) after HTA at 1360°C for 6 hr. A SiO_2 cap has been removed [29].

FIGURE 3.24   A set of cross-sectional TEM micrographs showing the existence of an optimum dose for the formation of a continuous BOX layer. The images (a) to (f) are from samples implanted with 1.5, 2.0, 2.5, 3.5, 4.5 and 5.0×10^{17} O^+ cm^{-2} at 65 keV. All samples were annealed at 1350°C for 6 hr in $Ar(O_2)$ with a SiO_2 cap [65].

forming a planar, homogeneous BOX with $t_{BOX} = 72$ nm and $T_{Si} = 146$ nm.

FIGURE 3.24 due to Jiao et al. [65] shows the dose dependence of the evolution of the BOX over the dose range 1.5×10^{17} O^+ cm^{-2} to 5.0×10^{17} O^+ cm^{-2} at an energy of 65 keV. All samples were annealed at 1350°C for 6 hr in Ar (O_2) with a TEOS silicon dioxide cap. The two lower dose samples contain a few relatively large amorphous SiO_2 precipitates with their major axis

41

TABLE 3.1 Optimum doses to form a thin continuous BOX layer containing no silicon islands. The samples were implanted in an IBIS-1000 machine and annealed in an $Ar + O_2$ ambient.

Energy (keV)	Dose (cm^{-2})	t_{BOX} (nm)	t_{Si} (nm)	Reference
65	2.0×10^{17}	50	<90	[76]
100	3.0×10^{17}	75	170	[70]

lying in the (100) plane of the wafer. The sample implanted with 2.5×10^{17} $O^+ cm^{-2}$ exhibits a continuous, homogeneous BOX with no entrapped silicon islands whilst samples implanted with higher doses have numerous faceted silicon islands in the BOX, again showing a memory effect as their major axis lies in the plane of the wafer. The sample implanted with a dose of 5×10^{17} $O^+ cm^{-2}$ has a predominance of large silicon islands near the lower BOX/Si interface similar to the sample shown in FIGURE 3.21. From these and other results the authors conclude that the as-implanted atomic oxygen profile plays a crucial role in the development of the BOX layer during the HTA. They confirm that there is an optimum dose ("Izumi window") at each implantation energy for the formation of a continuous BOX with no silicon islands. The authors report that at 65 keV, under their processing conditions, the optimum dose is 2.0×10^{17} $O^+ cm^{-2}$ without a deposited cap and 2.5×10^{17} $O^+ cm^{-2}$ when annealed with a silicon dioxide cap.

Optimum doses and measured layer thickness of thin SIMOX structures after an HTA without a cap have been taken from the literature for substrates implanted in an IBIS-1000 implanter [75] (see Section 3.5). Details are listed in TABLE 3.1.

Chen and co-workers [77, 78] have undertaken a systematic investigation of the energy dependence of the optimum dose to form a thin continuous BOX with a minimum density of entrapped silicon islands. The substrates were implanted at energies between 75 keV and 150 keV at a temperature of 680°C. The subsequent HTA was typically >1300°C for 5 hr in $Ar + x\%$ O_2 ($x < 3\%$) but no other special processing conditions or layer thicknesses were reported. This group found that a continuous BOX layer could be formed by an appropriate choice of O^+ dose, as shown in FIGURE 3.25 [77]. Also included in the figure are data taken from the literature, as identified in the caption. It is noted that the data from Ogura [73, 79] show the lowest value for the optimum dose to form a continuous thin BOX beneath a thick silicon overlayer (t_{Si}), using an implantation energy of 180 keV. Recently reported values of the optimum dose are listed in TABLE 3.2.

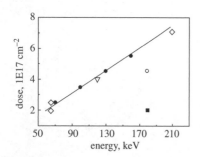

FIGURE 3.25   A plot of a series of good dose-energy matches for the formation of high quality low-dose SIMOX [77]. Key: ● [77], ○ [30], ■ [83], ▽ [80] and ◇ [65].

TABLE 3.2 Energy dependence of the optimum O^+ dose to achieve a continuous thin BOX with a minimum density of silicon inclusions after an HTA.

Energy (keV)	70	100	120	130	160	180
Dose $(cm^{-2}) \times 10^{17}$	2.5	3.5	4.0	4.5	5.5	4 ± 2
Reference	[77]	[77]	[80]	[77]	[77]	[73, 81]

In a recent series of papers [32, 33, 79] concerning the HTA of low-dose SIMOX materials, Ogura has developed the theme of controlling the evolution of the oxide precipitates, hence the microstructure of the BOX, by adjusting the anneal ambient and temperature ramp rate [68]. This is essentially an extension of the physics controlling the ITOX process: see below [66]. By so doing he is able to widen the "Izumi window" at a given O^+ implantation energy. For example, at an energy of 180 keV he finds the conventional dose window to be between 3.5×10^{17} O^+ cm^{-2} and 4.5×10^{17} O^+ cm^{-2} (viz. $4.0 \pm 0.5 \times 10^{17}$ O^+ cm^{-2}) with an anneal at 1350°C for 4 hr in $Ar + 0.5\%$ O_2 using a ramp rate of 20°C min^{-1}. By reducing the ramp rate to 0.02°C min^{-1} between 1000°C and 1340°C and with an ambient of $Ar + 1\%$ O_2 the dose window is widened to have upper and lower values of 2.0×10^{17} O^+ cm^{-2} and 6.0×10^{17} O^+ cm^{-2} (viz. $4.0 \pm 2.0 \times 10^{17}$ O^+ cm^{-2}), respectively. The expectation is that by judicious adjustments of the O^+ dose, annealing ambient and ramp rate it will be possible to simplify SIMOX processing and, thus, lower the commercial cost of thin SIMOX substrates.

ITOX process

Internal oxidation (ITOX) [31] used in the fabrication of thin film SIMOX substrates has been described as an enhanced low-dose SIMOX process based upon high temperature oxidation [66]. The process involves a second HTA ($\approx$1300°C) in a strongly oxidising ambient, $Ar + x\%$ O_2 with $10\% < x < 70\%$. Under these conditions the concentration of interstitial oxygen at the free surface will be higher than the oxygen solubility in silicon and the concentration gradient will drive oxygen to the BOX where internal oxidation will occur, predominantly at the upper Si/SiO_2 interface. As a consequence the thickness of the BOX (t_{BOX}) will increase whilst the thickness of the silicon overlayer (t_{Si}) will be reduced due to silicon consumption both by the ITOX layer and by growth of a sacrificial thermal oxide at the free surface. This is shown schematically in FIGURE 3.26(a) where t_{OX} and t_{ITOX} are defined. Cross-section TEM micrographs in FIGURE 3.26(b) contrast the

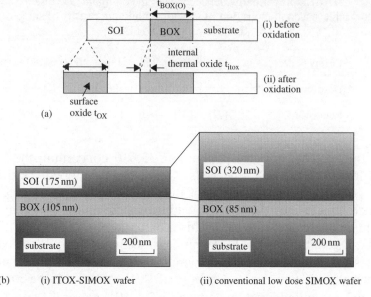

FIGURE 3.26 (a) Schematic showing the concept of the ITOX process (i) before and (ii) after oxidation. (b) Cross-sectional TEM micrographs showing (i) an ITOX-SIMOX structure (t_{BOX} = 105 nm) and (ii) a conventional low-dose SIMOX (t_{BOX} = 85 nm) [66].

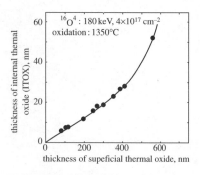

FIGURE 3.27 ITOX induced buried thermal oxide versus the surface thermal oxide grown on a SIMOX sample during an anneal at 1350°C [31].

structure of a conventional low-dose SIMOX substrate with the substrate after completion of the ITOX process [66].

Experiments show that the growth rate of the ITOX layer can be controlled in a predictable manner by adjusting the anneal temperature and time and oxygen concentration. FIGURE 3.27 shows the relative thicknesses of the ITOX and thermal (surface) oxides in a low-dose SIMOX structure (4×10^{17} O^+ cm^{-2} at 180 keV) during oxidation at 1350°C [84]. Thus the ITOX process leads to the growth of a thicker BOX without incurring the additional cost of a higher O^+ dose. The growth of the two oxide layers beneficially reduces the thickness of the silicon overlayer (t_{Si}), which is required for FD-CMOS devices [1].

Tachimori et al. [66] have shown that over the temperature range of 1000°C to 1350°C the thickness of the ITOX layer (t_{ITOX}) is given by:

$$t_{ITOX} \approx \frac{k}{0.90} \log_e \left(\frac{t_{BOX(O)}}{t_{BOX(O)} - 0.45 t_{OX}} \right)$$

where k = $2.25 \times 10^5 \exp(-(1.26\,eV/kT))$ [mm] and $t_{BOX(O)}$ is the original thickness of the BOX before oxidation. Under typical oxidation conditions for a 200 keV SIMOX substrate the practical maximum thickness of the additional ITOX layer is ≈50 nm [84].

After processing the BOX consists of two layers, namely an upper high quality thermal oxide formed by the ITOX process and a thicker oxide formed by IBS, which is in contact with the silicon substrate. Measurements by Anc et al. [85] show that ITOX-SIMOX materials have a reduced dislocation density in the silicon overlayer, a smoother Si/BOX interface and a reduced density of silicon inclusions in the BOX.

In summary, the ITOX process holds great promise as a route to reduce the cost of thin film SIMOX for FD-CMOS, achieve tighter control of layer thicknesses, and improve the integrity of the BOX, especially the voltage breakdown performance.

"Top-up" implant plus ITOX

The "top-up" process [84] is designed to enhance the conventional ITOX process [66], as described above, by introducing an additional layer of lattice defects near the Si/BOX interface, which increases the rate of internal oxidation. It entails an additional "top-up" implantation of a small dose of O^+ ions at ambient (room) temperature to create an excess population of point defects due to the collision cascades. It is reported that the oxidation rate is faster by a factor of two thus almost doubling the thickness of the ITOX layer [84]. When applied to thin film SIMOX materials it further enhances the integrity and thickness of the BOX resulting in devices showing even fewer electrical shorts and still further improved voltage breakdown performance.

3.4 NOVEL SIMOX PROCESSING AND STRUCTURES

The discussion so far has concerned planar whole wafer SOI structures comprising a single continuous BOX, as shown in FIGURE 3.1. These substrates are most frequently used for advanced CMOS device applications; however, new uses are currently being investigated thanks to possibilities opened through a better understanding of the SIMOX process, improved processing equipment and the improving quality and lower unit cost of SOI wafers. For completeness novel SIMOX structures currently being researched will be briefly discussed below, although some structures and mechanisms are reviewed in greater detail in other chapters of this volume.

3.4.1 SPIMOX

Separation by plasma implanted oxygen (SPIMOX) is the name given to SIMOX structures formed by plasma immersion ion

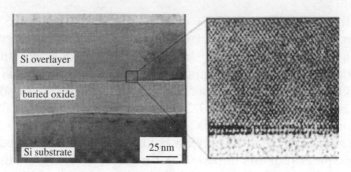

FIGURE 3.28   Cross-sectional TEM micrographs of a SPIMOX sample implanted at 60 keV with a dose of 2×10^{17} O^+ cm^{-2} and annealed at 800°C for 1 hr plus 1320°C for 2.5 hr [72].

implantation (PIII), in which the wafer is immersed in an oxygen plasma [86], as described in Section 3.5. A major advantage of PIII over conventional ion implantation is that it offers high dose rate implantation, hence shorter implantation times and lower processing costs, and is compatible with large area processing [72]. FIGURE 3.28 shows XTEM micrographs from an annealed SPIMOX structure formed by implantation of 2×10^{17} O^+ cm^{-2} at a bias voltage of 60 kV and annealed at 800°C for 1 hr followed by 1320°C for 2.5 hr. The BOX layer shows some non-uniformity although the Si/BOX interfaces are abrupt, as shown in the high resolution image [72].

FIGURES 3.29(e) and (f) show XTEM micrographs from as-implanted and annealed SPIMOX material formed by the concurrent implantation of $O^+ + O_2^+$ ions with doses of 1.8×10^{17} cm^{-2} [86]. The bias voltage was set at 50 kV and, as the $O^+ + O_2^+$ ions were implanted concurrently, the oxygen atoms had kinetic energies of 50 keV and 25 keV, respectively. The sample was capped with Si_3N_4 and annealed at 1250°C for 2 hr in a flowing nitrogen ambient. The as-implanted sample (FIGURE 3.29(e)) has a broad highly damaged region reminiscent of conventional SIMOX samples. The annealed sample (FIGURE 3.29(f)) contains two BOX layers with very wavy interfaces that are highly non-uniform in thickness, highlighting the need for further process development. However, in the present context, the inherent strength of PIII is the ability to concurrently implant two different ion species and directly synthesise double BOX structures without incurring the additional implantation overheads of two separate implantation steps.

Chen et al. [78] have recently reported improved quality SPIMOX materials which have been fabricated using a water plasma as the source of oxygen [72], which ensures the most abundant ion species (H_2O^+, HO^+ and O^+) all have similar

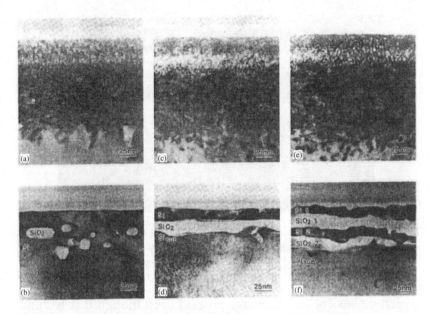

FIGURE 3.29   Cross-sectional TEM micrographs of SPIMOX samples implanted at 50 keV with (a, b) 1×10^{17} O^+ cm^{-2}, (c, d) 3×10^{17} O^+ cm^{-2} and (e, f) $O^+ + O_2^+$ to a dose of 1.8×10^{17} O^+ cm^{-2}. (a, c and e) as-implanted and (b, d and f) after HTA at 1250°C for 2 hr. Note the presence of a double BOX in (f) [86].

atomic masses and, thus, similar penetration depths into the silicon target. They identified an optimum ion dose-energy window as 5.5×10^{17} cm^{-2} at 90 keV and report structures with smooth, abrupt Si/BOX interfaces after an anneal at 1320°C for 5 hr in Ar + 1% O_2.

3.4.2   Multilayers

The ability to synthesise a BOX multilayer structure was demonstrated in 1987 by Hemment et al. [87] who implanted doses of 9×10^{17} O^+ cm^{-2} at energies of 350 keV and 200 keV at an implantation temperature of $\sim$550°C. The wafers were annealed at 1405°C for 30 minutes. FIGURE 3.30 shows XTEM micrographs from the resulting structure consisting of a 230 nm thick single crystal silicon overlayer, a 200 nm BOX, 260 nm of single crystal silicon and a second 180 nm BOX layer. The interfaces are abrupt but with pronounced waviness at the intermediate silicon layer which contained a high density of dislocations and thin rod-like defects [87]. No attempt was made to optimise the processing and only a double BOX structure was fabricated. Unfortunately this route to form multilayer structures involves serial implantations and high O^+ ion energies to achieve the overall layer thickness

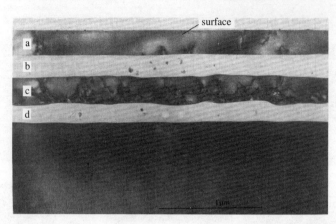

FIGURE 3.30   Cross-sectional XTEM micrograph showing a double BOX structure in a wafer implanted with doses of 0.9×10^{18} O^+ cm^{-2} at 350 keV and 200 keV and annealed at 1405°C for thirty minutes [87].

which makes for an expensive process even though it is technically feasible. In an attempt to reduce this high cost Chen et al. [88] have demonstrated that thin multi BOX structures can be formed by implanting low-doses of O^+ ions using a combination of matched low-doses and energies. Ogura et al. [79] have also successfully fabricated a thin film double BOX structure by implanting a wafer with a low O^+ dose and annealing in an oxygen-rich ambient using a very slow ramp rate. The resulting structure is shown in FIGURE 3.33(c).

An alternative approach is to employ plasma immersion ion implantation (PIII) as reported by Liu et al. [86] during the mid 1990s. They exploited the strengths of PIII (Section 3.5) and concurrently implanted two oxygen ion species (O^+ and O_2^+) by using an oxygen feeder gas and setting the plasma conditions to give the appropriate abundances of O^+ and O_2^+. The bias voltage was set at 50 kV and oxygen was implanted at 50 keV and 25 keV, respectively, due to the different atomic masses of the two ion species. The resulting double BOX structure is shown in FIGURE 3.29(f).

3.4.3   Patterned SIMOX

The commercialisation of SIMOX based integrated circuits is now a reality and it has been suggested that future improvements of circuit performance will come from the fabrication of circuits on patterned SOI substrates [84]. Several different routes to form patterned structures have been explored ranging from selected area O^+ implantations on a lateral scale of one micron with the aim of vertically isolating individual CMOS source and drain junctions from the substrate [89], to the fabrication of complete circuit components on adjacent SOI and bulk silicon regions.

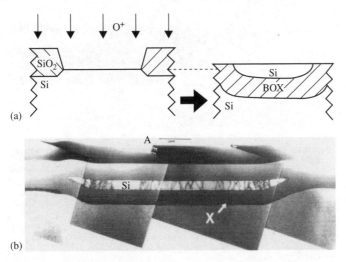

(a)

(b)

FIGURE 3.31   (a) Schematic illustrating the LOCOS geometry employed to define SIMOX islands and realise a planar patterned structure [90]. (b) Cross-sectional TEM micrograph of a dielectrically isolated silicon device island (TDI). The BOX is labelled "X". A high density of threading dislocations is evident in the silicon overlayer [91].

Early O^+ implantations into patterned wafers were reported by Bussmann et al. [90] in the context of achieving total device isolation (TDI). He employed a LOCOS masking system and optimised the thickness of the oxide in order to exploit the higher sputter yield of SiO_2 so that for a given O^+ ion dose and energy, he could achieve a planarised final structure. The method is shown schematically in FIGURE 3.31(a). Largely due to the use of a high O^+ dose the isolated silicon islands were highly defective with entrapped silicon precipitates in the vicinity of the mask edge. More recently Iyer et al. [84] have reported the successful realisation of low defect density patterned SIMOX with a nominal BOX thickness of 80 nm, as shown in FIGURE 3.32(b). Chen et al. [92] have realised patterned SOI structures using matched O^+ doses and energies and report successful operation of fabricated CMOS devices. They further report that the implantation of very low doses (LII) reduces the density of window edge defects thus opening the way for commercial exploitation [93].

3.4.4   SIMOX without O^+ implantation?

Recent years have seen very detailed investigations of the evolution of the BOX layer during HTA with special attention being given to the role of the anneal temperature, ramp rate, time-temperature profile and the annealing ambient. The outcome is an ability to engineer the BOX in terms of thickness, location and microstructure by selectively stimulating the growth of oxide precipitates

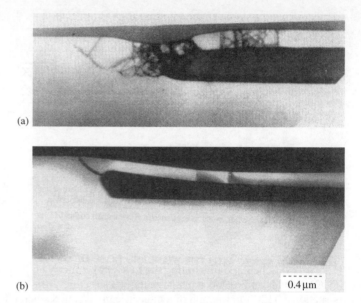

(a)

(b)   0.4 μm

FIGURE 3.32   Cross-sectional TEM micrographs showing the mask edge of patterned SIMOX wafers. (a) High dose sample (t_{BOX} = 370 nm). (b) Low-dose (t_{BOX} = 80 nm). There is a significant reduction in the density of defects at the mask edge in the low-dose sample [72].

(Ostwald ripening [56]). In Section 3.3.2 it was shown that a high quality BOX can be achieved at the peak of the implanted oxygen profile. Now it is recognised that a thin BOX layer can be forced to evolve at the peak of the damage profile [81]. The anneal schedules that are required include an HTA with a slow ramp rate, especially over the temperature range 1200°C to 1340°C, and an oxygen rich anneal ambient [94] which facilitates oxygen in diffusion and provides time for the growth and hence stabilisation of small oxide precipitates located at the peak of the damage profile [68].

FIGURE 3.33, due to Ogura and Ono [83], shows XTEM micrographs from similarly annealed samples (1340°C, 0.1°C min^{-1} and 14% O_2) implanted at ~600°C with O^+ ions at an energy of 180 keV but different oxygen doses, namely (a) 2×10^{17} O^+ cm^{-2}, (b) 3×10^{17} O^+ cm^{-2} and (c) 4×10^{17} O^+ cm^{-2}. In the figure the depths of the peaks of the damage and atomic oxygen profiles (see FIGURE 3.3) are identified as D_p and R_p, respectively. For each dose a similar continuous BOX layer has formed at D_p whilst at R_p only a layer of oxide precipitates exists in the lowest dose sample. The density of the precipitates is higher in the sample implanted with 3×10^{17} O^+ cm^{-2} and evolves into a continuous BOX for the highest dose. From these and other supporting experiments [83] Ogura finds that double BOX structures, shown in FIGURE 3.33(c), can be formed reproducibly where the BOX at depth R_p is associated with the implanted O^+ whilst the growth of the BOX closer to the surface is due to in-diffusion of oxygen from the anneal ambient.

Based upon the above observations Ogura [33] has proposed and demonstrated that SOI structures can be formed when the nucleation sites for precipitate growth are lattice defects created during light ion implantation, specifically H^+ and He^+, and the supply of oxygen is not provided by O^+ ion implantation but by diffusion from an oxygen rich ambient containing between 6% and 20% oxygen. FIGURE 3.34 shows XTEM micrographs from He^+ irradiated and annealed samples. The processing conditions included implantation of 45 keV He^+ ions to doses of (a) 2×10^{17} He^+ cm^{-2} and (b) 4×10^{17} He^+ cm^{-2} followed by an HTA at 1340°C with a slow ramp rate and a quoted oxygen content between 6% and 20% [33]. The sample implanted with the lower dose of He^+ ions contained only faceted oxide precipitates at the peak of the damage profile; however, a continuous BOX layer forms in the sample implanted with 4×10^{17} He^+ cm^{-2}. Thus, by a blending of the light ion implantation of Unibond© [95] and the very high temperature anneal of SIMOX, Ogura has realised SOI structures that potentially are less expensive to manufacture and contain fewer crystallographic defects. Only time will tell. ...

3.5 SIMOX TECHNOLOGY AND ENGINEERING

3.5.1 Background

The original investigations of IBS referred to in Section 3.1, which culminated in the successful development of SIMOX and related technologies, were initiated during the 1950s and 1960s, long before the development of modern high current ion implanters. It is ironic that high current mass separators were also not available at that time as, during the 1940s in a different context, a technology (Calutrons) had been developed to generate and transport high current U^+ ion beams to produce ^{235}U enriched uranium for the World War II Manhattan Project [96]. Consequently the original IBS experiments were carried out on "home made" implanters with low beam currents typically of order 10 μA. The high doses, in the range 10^{17} O^+ cm^{-2} to 10^{18} O^+ cm^{-2}, could only be achieved by reducing the implanted area to one or, at best, a few cm^2 and accepting long implantation times of many hours or even days. Fortunately this could be justified in the academic research environment where many of the experiments were carried out.

Early experiments soon highlighted the importance of dynamic annealing during the O^+ implantation but, in the absence of heated

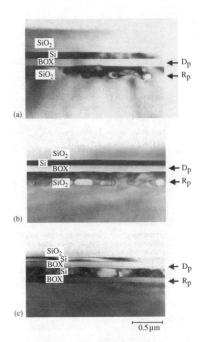

FIGURE 3.33 SOI structures realised in samples implanted with (a) 2×10^{17} O^+ cm^{-2}, (b) 3×10^{17} O^+ cm^{-2} and (c) 4×10^{17} O^+ cm^{-2} at 180 keV after HTA at 1340°C in Ar/O_2. In (a) and (b) a continuous BOX is formed at the peak of the damage profile (D_p) whilst in (c) a double BOX is formed with layers at the damage peak (D_p) and at the peak of the atomic oxygen profile (R_p) [83].

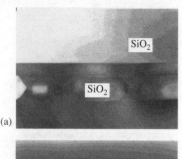

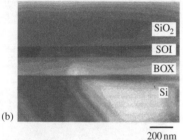

FIGURE 3.34 Cross-sectional TEM micrographs from samples implanted with
(a) 2×10^{17} He$^+$ cm^{-2} and
(b) 4×10^{17} He$^+$ cm^{-2} at 45 keV after HTA at 1340°C for 4 hr in an Ar/O$_2$ ambient. (b) Shows the presence of a continuous BOX where the source of oxygen is the annealing ambient [33].

chucks upon which to mount the wafers, the power (P = IV) carried by the ion beam was used to heat the target and achieve the desired annealing temperature [89]. This is possible because a silicon wafer in vacuum, with no conductive cooling, will achieve an equilibrium temperature of 400°C to 500°C at a modest power loading of 1 W cm^{-2} (say, 5 μA cm^{-2} at 200 keV). The grave disadvantage of beam heating, however, is that the ion flux (beam current) and wafer temperature cannot be independently controlled. Also, O$^+$ ion channelling and accumulation of lattice damage will occur during the initial stages of the implantation, whilst the wafer is heating up to the required equilibrium temperature [42]. For these reasons volume production of high quality SIMOX wafers for commercial and military applications became possible only during the late 1980s with the arrival of dedicated high current commercial O$^+$ implanters with end stations equipped to pre-heat and provide background heating of large batches of wafers, thereby achieving consistency in the microstructure of the implanted wafers.

3.5.2 Machine technologies

The pioneering experiments by Izumi and colleagues at NTT [27] used an Extrion 200-20A implanter, a single wafer, medium current machine with electrostatic beam scanning. The ion beam current was enhanced by NTT who modified the PIG cold cathode ion source, installing internal titanium components in order to use a pure oxygen source gas [97]. They achieved a stable ^{16}O$^+$ beam current of up to 100 μA. Elsewhere a number of industrial and academic groups also used Extrion machines or derivatives.

The late 1970s saw the development of mechanical scanning systems using a spinning disk [98] which facilitated exploitation of the high current (several mA) Freeman source [99]. Subsequently dopant implanters, such as the Easton NV10-160, were operating with a nominal 10 mA rating. During 1984 examples of this machine were modified to include a heated end station and oxygen tolerant components in the ion source to permit mA currents of O$^+$ ions to be generated. During 1984 machines were delivered to Texas Instruments and IBM and run with O$^+$ and N$^+$ beams to investigate SIMOX and SIMNI technologies [89].

Arguably the most important step in the SIMOX odyssey was the corporate decision by NTT to fund the development of a new dedicated 100 mA O$^+$ implanter with a design capacity to process batches of twenty four 4″ wafers in about two hours [97]. Such a machine was designed to test the viability of SIMOX technology in terms of engineering, material science and commercial issues. Consequently, during 1984 NTT and the Eaton Corporation

FIGURE 3.35   Eaton NV-200 end station. Wafers are mounted on the inside of a drum and the ion beam is deflected below the horizontal. Manual wafer loading is through the port in the top of the vacuum chamber [75].

entered into a partnership to design and build the first dedicated SIMOX implanter, where the equipment vendor was responsible for the hardware and software including machine operation whilst NTT provided the design for a new ecr ion source [100], based upon research and development at Hitachi [89], and also carried out machine evaluations. The machine (NV-200) was installed at NTT and SIMOX wafer production started during 1986. FIGURE 3.35 is a photograph of the end station of an NV-200 showing that the O^+ beam is directed down onto the wafers mounted on the inside surface of a drum [75]. Rotation of the drum about a vertical axis distributed the beam power, of up to 20 kW ($P = IV$), over the batch of wafers and ensured lateral dose uniformity. Vertical dose uniformity was achieved by using a multi-aperture ion source extraction electrode with the overlapping beamlets focused to form a fanned beam [75].

Six Eaton NV-200 machines were built and delivered to NTT, Japan, Nippon Steel, Japan (2), LETI, France, Fraunhofer Institute, Germany, and Texas Instruments, Dallas, USA. The prototype machine was subsequently used by Eaton to produce SIMOX material to meet the market demand for SOI substrates [100].

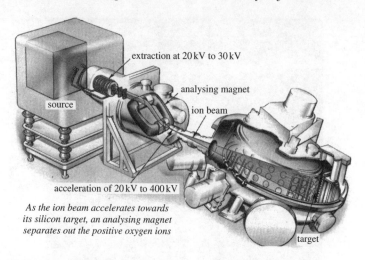

extraction at 20 kV to 30 kV

analysing magnet

source

ion beam

acceleration of 20 kV to 400 kV

*As the ion beam accelerates towards
its silicon target, an analysing magnet
separates out the positive oxygen ions*

target

FIGURE 3.36   Schematic of the OXIS machine showing the ion source, magnetic spectrometer and end station containing a rotating drum assembly [101].

This same period (mid-1980s) saw the development in the UK of an alternative design of O^+ implanter for SIMOX technology, funded by the UK Government under the Alvey Programme [101]. The project was undertaken in a four way partnership involving an equipment vendor (VG Semicon Ltd.), government laboratories (AERE Harwell and UKAEA Culham), the industrial customer base (GEC, Plessey and BTRL) and the academic community for technical input. FIGURE 3.36 is a schematic of the resulting "OXIS" machine which incorporated novel designs for various subsystems including the ion source, ion optics, mass analysis and beam transport. The original specification included beam currents of up to 200 mA, energy adjustable between 30 keV and 400 keV and an end station to accept large batches of 4″ wafers in rows of four with the facility to pre-heat the wafers up to 600°C. Unfortunately OXIS was a second generation machine being built without design and operational experience from a first generation machine. Even with an estimated (1986) purchase price of £4 M ($6 M) for the first machine it proved to be too expensive to complete the development and the project was terminated in 1987 [102].

In 1987 Guerra leased the prototype NV-200 from Eaton and founded IBIS Technology Corp as a SIMOX wafer vendor [100]. The company ramped up commercial production of SIMOX wafers and in so doing gained valuable experience and knowledge of the foibles of SIMOX wafer processing. Under pressure of market demand for substrates the company undertook the design of a second generation machine (IBIS-1000) which would be capable of volume production of improved SIMOX substrates at lower cost. The known weaknesses of the NV-200 were addressed as follows [100].

- Metal and carbon contamination – all beamline components visible to the O^+ beam were protected with silicon tiles. Graphite wafer holders were introduced with silicon pins to retain the wafers.
- Lateral dose uniformity – wafers were mounted on a spinning wheel rather than a drum. Hybrid scanning was employed with the addition of 150 Hz magnetic scanning where the scan rate could be adjusted. These changes served both to reduce ion channelling and to enhance lateral dose uniformity.
- Dislocation and stacking fault densities – an enhanced, forced air cooled quartz halogen lamp system in the endstation provided better control of wafer warm up and temperature during implantation which also facilitated multiple energy implantations and O^+ ion energy profiling.

FIGURE 3.37 shows the interior furniture of the IBIS-1000 endstation [100]. The engineering enhancements gave greater control over the quality of the processed wafers and enabled the targets set in the International Technology Roadmap for Semiconductors (NTRS) [34] to be adequately achieved or surpassed. Wafer specifications published by IBIS Technology include: metal contamination $<10^{11}$ atom cm^{-2}, dose uniformity $<\pm1\%$ and surface roughness 3.0 Å which were maintained during volume production of SIMOX substrates [100]. Recently, the Shanghai Institute of Microsystems and Information Technology, China, has announced the purchase of an IBIS-1000 implanter for the production of SIMOX substrates [103].

The engineering and commercial challenge for the SIMOX vendors is to meet the increasing demand for SIMOX substrates

FIGURE 3.37   End station of the IBIS-1000 showing wafer holders and the hub [100].

whilst still improving substrate quality, achieving lower unit costs and meeting the current targets of the ITRS. In order to address these issues IBIS Technology designed and built a new high current SIMOX implanter (i2000) optimised for 200 mm and 300 mm wafer processing [100]. This machine incorporates further engineering improvements and has a specification that includes: energy 40 keV to 240 keV, and O^+ beam current up to 80 mA using an ecr source with oxygen feeder gas and implantation temperature 400°C to 600°C. The beam is now collimated magnetically to ensure a constant angle of incidence during the scan cycle and only illuminates silicon components in the beamline and endstation. Wafer handling is fully automatic being compatible with FOUP and SMIFF load ports with a robot operating in the high vacuum chamber. Measured performance figures from this new machine include lateral layer thickness variations across a wafer of $t_{BOX} < \pm1\%$ and $t_{Si} \pm 50$ Å and $< \pm100$ Å batch to batch [104].

Hitachi Ltd. have twenty to thirty years of design and manufacturing experience of ion implantation equipment, including ecr ion sources, for commercial dopant implanters (IP815) [105]. Indeed their expertise contributed to the development of the ecr source for the original Eaton NV-200 machine [106]. Operational experience has shown this type of source to be most suitable for high current O^+ machines. During the 1990s the company developed a dedicated 200 keV, 100 mA SIMOX machine (UI-5000) incorporating an ecr high current source [100]. Dose uniformity was achieved by mechanically scanning (laterally rocking) the rotation disk (500 rpm) upon which thirteen 200 mm wafers or seventeen 150 mm wafers could be loaded. Lamp heaters provided independent control of the wafer temperature during implantation, typically at 550°C. The original design was subsequently updated to meet new ITRS targets and become the UI-6000 [105]. A similar scanning system is used but with the rotating disk (350 rpm) accepting twelve 300 mm wafers or eighteen 200 mm wafers as shown in FIGURE 3.38. The beam current rating is 100 mA and energy 40 keV to 240 keV with implantation temperature adjustable between 500°C and 650°C. The measured performance of the UI-6000 includes: dose uniformity $< \pm1.4\%$, uniformity $t_{BOX} \pm 1$ nm, metal contamination $\approx 10^{10}$ atoms cm^{-2} and particle count <0.5 cm^{-2} at >0.2 μm diameter [105]. Machines have been installed in Japan and also one UI-5000 has been sold to the USA.

FIGURE 3.38 The rotation wheel of the Hitachi UI-6000 showing the wafer holders and hub [100].

3.5.3 Plasma immersion ion implantation (PIII)

SIMOX technology was developed using beamline implanters that are derivatives of the original mass separators; however, alternative plasma implantation techniques are attractive, especially plasma

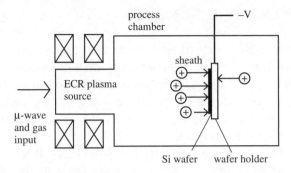

FIGURE 3.39   Conceptual schematic of a PIII system used to fabricate SPIMOX materials [107].

immersion ion implantation (PIII) which has been applied to semiconductor processing since the late 1980s [87]. Conceptually the technique is very simple, as shown in FIGURE 3.39, which is a schematic of a PIII system for the fabrication of separation by plasma implanted oxygen (SPIMOX) substrates [107]. Here the silicon wafer is mounted on a heated sample holder around which an oxygen plasma is established. A fast negative voltage pulse, typically of a few tens of kV, is applied to the holder causing the electrons in the immediate vicinity to be repelled leaving a net positively charged (O^+) ion sheath. The positive ions are accelerated towards the negatively biased wafer thus gaining kinetic energy causing them to be implanted into the silicon target. Upon removing the negative bias the sheath will collapse and recovery occurs prior to the next pulse [72]. As it is customary to operate under conditions of minimal ion scattering (collisions) the maximum implantation energy is defined approximately by the applied bias voltage.

The benefits of PIII include mechanical simplicity and short implantation times as high ion fluxes can be achieved ($\approx 10^{16}$ ions cm^{-2} s^{-1}) thanks to a high plasma density of 10^9 ions cm^{-3} to 10^{11} ions cm^{-3}. The technique can be scaled up for large area targets and has the facility for conformal implantation of targets with a non-planar topography. However, there are inherent problems with the process, which include:

- multiple ion and chemical species may be implanted – as no momentum analysis;
- energy spread – as the ions have different path lengths;
- poor control of dose and dose uniformity;
- maximum energy typically $\ll 100$ keV.

Despite these issues concerning process control, SOI wafers have been successfully produced, as shown in FIGURE 3.28.

3.6 CONCLUDING REMARKS

Silicon on insulator substrates have been a potential replacement for bulk silicon for "next generation CMOS circuits" for two decades; however, bulk silicon has been able to hold off the challenge thanks to advances in processing and device architectures. However, now that high quality, lower cost SOI substrates are commercially available and as the ITRS [34] includes targets for SOI materials the long-term prospect for SIMOX is much improved. Commercial SIMOX ICs are on the market and the shift from thick to thin film SOI substrates is now gaining pace. It is noted that recent advances in processing of thin BOX structures has been achieved by careful matching of the O^+ ion energy and dose and by optimising the anneal ambient and ramp rate. These developments pave the way to higher quality substrates and still lower processing costs.

It is noteworthy that the authors [108] of a recent (2002) critical appraisal of the three principal SOI materials technologies (SIMOX, Unibond [95] and ELTRAN® [82]) have identified SIMOX as their preferred technology as, they argue, the substrates are less expensive, manufacture is by an inherently simple process and the technology already has widespread industrial acceptance.

The physics and chemistry of the SIMOX process are largely understood and volume manufacture of high quality SIMOX/SOI substrates is now becoming a reality with the wafer vendors optimistic that the targets specified in the ITRS for the next generation of CMOS SOI circuits can be adequately met. However, the real challenge for SIMOX is in the commercial arena – can production be ramped up quickly to meet market demand with a wafer cost that industry can accept and without compromising substrate quality? If the answer is "yes" then the next step is to secure the investment required to build and install an adequate number of SIMOX implanters to meet the global demand. However this depends upon the acceptance by the circuit design community of FD-CMOS circuits on a thin BOX layer as this is the viable route to lower cost through smaller O^+ doses and shorter implantation times.

REFERENCES

[1] J.-P. Colinge [*Silicon-on-Insulator Technology* (Kluwer Academic Publishers, 1997) 2nd edition]

[2] R. Ohl [*Bell Syst. Tech. J. (USA)* vol.31 (1952) p.104]

[3] W.D. Coussins [*Proc. Phys. Soc. Lond. B (UK)* vol.68 (1955) p.213]

[4] M.L. Smith (Ed.) [*Electromagnetically Enriched Isotopes and Mass Spectroscopy* (Butterworth Press, London, 1956) p.100]

[5] W. Shockley [US Patent No. 2,787,564 (1957)]

[6] M.M. Brodov, V.A. Lepilin, I.B. Shestakov, A.L. Shakh-Budagov [*Sov. Phys.-Solid State (USA)* vol.3 (1961) p.195]

[7] M. Watanabe, A. Tooi [*Jpn. J. Appl. Phys. (Japan)* vol.5 (1966) p.737]

[8] P.V. Pavlov, E.V. Shitova [*Sov. Phys. (USA)* vol.12 (1967) p.11]

[9] V.M. Gusev et al. [*Radio Eng. Electron. Phys. (USA)* vol.16 (1971) p.1357]

[10] C.R. Fritzche, W. Rothemund [*J. Electrochem. Soc. (USA)* (1971) p.1243]

[11] J.H. Freeman, G.A. Gard, D.J. Mazey, J.H. Stephen, F.B. Whiting [*European Conf. on Ion Implantation* Reading, 1970 (Peregrinus, Hitchin, Herts, 1974) p.74]

[12] H.M. Naguib, R. Kelly [*Radiat. Eff. (UK)* vol.25 (1975) p.1]

[13] R. Kelly [*Radiat. Eff. (UK)* vol.64 (1982) p.205]

[14] P.L.F. Hemment [*Mater. Res. Soc. Symp. Proc. (USA)* vol.53 (1986) p.207–21]

[15] G.H. Schwuttke, K. Brack, E.D. Gardner, H.M. DeAngelis [*Proc. Conf. on Radiation Effects in Semiconductors* Ed. F.L. Vook (Plenum Press, New York, 1968) p.406]

[16] U. Bonse, M. Hart, G.H. Schwuttke [*Phys. Status Solidi (Germany)* vol.33 (1969) p.361]

[17] B. Williams [private communication]

[18] D. Dylewski, M.C. Joshi [*Thin Solid Films (Switzerland)* vol.35 (1976) p.327]

[19] D. Dylewski, M.C. Joshi [*Thin Solid Films (Switzerland)* vol.35 (1976) p.241]

[20] D. Dylewski, M.C. Joshi [*Thin Solid Films (Switzerland)* vol.42 (1977) p.227]

[21] M.H. Badawi, K.V. Anand [*J. Phys. D (UK)* vol.10 (1977) p.1931]

[22] R.J. Dexter, S.B. Watelski, S.T. Picreaux [*Appl. Phys. Lett. (USA)* vol.23 (1973) p.455]

[23] K. Das et al. [*Inst. Phys. Conf. Ser. (UK)* vol.60 (1981) p.307]

[24] S. Gill, I.H. Wilson [*Thin Solid Films (Switzerland)* vol.55 (1978) p.435]

[25] V.P. Astakhov, L.Ya. Konyushenko, V.M. Pyatakov [*Inorg. Mater. (USA)* vol.13 (1978) p.43]

[26] K.I. Kirov, E.D. Atanasova, S.P. Alexandrova, B.G. Amov, A.E. Djakov [*Thin Solid Films (Switzerland)* vol.48 (1978) p.187]

[27] K. Izumi, M. Doken, H. Ariyoshi [*Electron. Lett. (UK)* vol.14 (1978) p.593]

[28] C. Jaussaud, J. Stoemenos, J. Margail, M. Dupuy, B. Blanchard, M. Bruel [*Appl. Phys. Lett. (USA)* vol.46 (1985) p.1064]

[29] Y. Li et al. [*Appl. Phys. Lett. (USA)* vol.63 no.20 (1993) p.2812–4]

[30] S. Nakashima, K. Izumi [*J. Mater. Res. (USA)* vol.8 (1993) p.523]

[31] S. Nakashima, T. Katayama, Y. Miyamura, A. Matsuzaki, M. Imai, K. Izumi, and N. Ohwada [*Proc. IEEE SOI Conf.* (1994) p.71]

[32] A. Ogura [*Appl. Phys. Lett. (USA)* vol.74 no.15 (1999) p.2188–90]

[33] A. Ogura [*Jpn. J. Appl. Phys. (Japan)* vol.40 (2001) p.L1075–7]

[34] International Technology Roadmap for Semiconductors (ITRS) can be found at http://public.itrs.net

[35] E. Briggs, M.P. Seah (Eds.) [*Practical Surface Analysis Vol.2 'Ion and Neutral Spectroscopy'* (John Wiley, 1992) Appendix 3]

[36] J.F. Ziegler [in *Ion Implantation Science and Technology* Ed. J.F. Ziegler (Ion Implantation Technology Co, Maryland, USA, 2000) ISBN 0-9654207-0-1]

[37] J.F. Ziegler, J.P. Biersack, U. Littmark [in *The Stopping and Ranges of Ions in Solids* (Pergamon Press, New York, 1985) vol.1] TRIM may be obtained at http://www.research.ibm.com/ionbeams/

[38] I.H. Wilson [in *Ion Beam Modification of Insulators* Eds. P. Mazzoldi, G. Arnold (Elsevier Science Publishers BV, 1987) p.245–300]

[39] K.S. Jones, S. Prussin, E.R. Weber [*Appl. Phys. A (Germany)* vol.45 (1988) p.1]

[40] H.J. Stein [*Electrochem. Soc. Proc. (USA)* (1986)]

[41] O.W. Holland, L. Xie, B. Nielsen, D.S. Zhou [*J. Electron. Mater. (USA)* vol.25 (1996) p.99]

[42] H. Ryssel, I. Ruge [*Ion Implantation* (Wiley Interscience, 1986)]; P.L.F. Hemment et al. [*Nucl. Instrum. Methods Phys. Res. B (Netherlands)* vol.37/38 (1989) p.766]

[43] P.L.F. Hemment et al. [*Nucl. Instrum. Methods Phys. Res. B (Netherlands)* vol.21 (1987) p.129–33]

[44] J.A. Kilner et al. [*Nucl. Instrum. Methods Phys. Res. B (Netherlands)* vol.7/8 (1985) p.293–8]

[45] F. Namavar, T.I. Budnick, F.H. Sanchez, H.C. Hayden [*Mater. Res. Soc. Symp. Proc. (USA)* vol.53 (1986) p.233]

[46] P. Scanlon et al. [*Mater. Res. Soc. Symp. Proc. (USA)* vol.107 (1988) p.141–5]

[47] C. Jaussaud, J. Stoemenos, J. Margail, A.M. Papon, M. Bruel [*Vacuum (UK)* vol.42 no.5/6 (1991) p.341–7]

[48] S. Mantl [*Mater. Sci. Rep. (Netherlands)* vol.8 no.1/2 (1992) p.1–95]

[49] A. Borghesi, A. Sassella, A. Stella [*J. Appl. Phys. (USA)* vol.77 no.9 (1995) p.4169–244]

[50] G.F. Cerefolini, S. Bertoni, L. Meda, C. Spaggiari [*Nucl. Instrum. Methods Phys. Res. B (Netherlands)* vol.84 (1994) p.234]

[51] A. Bourret, J. Thibault-Desseaux, D.N. Seidman [*J. Appl. Phys. (USA)* vol.55 no.4 (1984) p.825]

[52] C. Jaussaud, J. Stoemenos, J. Margail, A.M. Papon, M. Bruel [*Vacuum (UK)* vol.42 no.5/6 (1991) p.341–7]

[53] R.D. Doherty [in *Physical Metallurgy* Eds. R.W. Cahn, P. Haasen (North Holland, Amsterdam, 1983) p.933]

[54] K.J. Reeson et al. [*Microelectron. Eng. (Netherlands)* vol.8 (1988) p.163–86]

[55] J. Stoemenos [*Nucl. Instrum. Methods Phys. Res. B (Netherlands)* (1996)]; J. Stoemenos [*Microelectron. Eng. (Netherlands)* vol.22 (1993) p.307]

[56] R. Reiss, K.J. Heinig [*Nucl. Instrum. Methods Phys. Res. B (Netherlands)* vol.84 (1994) p.229–33]; Porter, Easterling [*Phase Transformations in Metals and Alloys*]

[57] R.D. Doherty [in *Physical Metallurgy* Eds. R.W. Cahn, P. Haasen (North Holland, Amsterdam, 1983) p.933]

[58] A. Borghesi, A. Sassella, A. Stella [*J. Appl. Phys. (USA)* vol.77 no.9 (1995) p.4169–244]

[59] L.C. Brown [*Scr. Metall. (USA)* vol.21 (1987) p.693]

[60] J. Margail, J.M. Lamura, J. Stoemenos, A.M. Papon [*Electrochem. Soc. Proc. (USA)* vol.92-13 (1992) p.407]

[61] S. Reiss, K.H. Heinig [*Nucl. Instrum. Methods Phys. Res. B (Netherlands)* vol.84 (1994) p.229–33] Y. Li et al. [*Nucl. Instrum. Methods Phys. Res. B (Netherlands)* vol.85 (1994) p.236]

[62] P.L.F. Hemment, R.F. Peart, M.D. Yao, K.G. Stephens [*Appl. Phys. Lett. (USA) vol.46 (1985) p.952*]

[63] G.K. Celler, P.L.F. Hemment, K.W. West, J.M. Gibson [*Appl. Phys. Lett. (USA) vol.48 no.8 (1986) p.532*]

[64] B. Johnson, Y. Tan, P. Anderson, S. Seraphin, M.J. Anc [*J. Electrochem. Soc. (USA) vol.148 no.2 (2001) p.663–7*]

[65] J. Jiao, B. Johnson, S. Seraphin, M.J. Anc, R.P. Dolan, B.F. Cordts [*Mater. Sci. Eng. B (Switzerland) vol.72 (2000) p.150–5*]

[66] M. Tachimori et al. [*Electrochem. Soc. Proc. (USA) vol.96-3 (1996) p.53–62*]

[67] S. Nakashima et al. [*J. Electrochem. Soc. (USA) vol.143 (1996) p.244*]

[68] Y. Tan, B. Johnson, S. Seraphin, J. Jiao, M.J. Anc, K.P. Allen [*J. Mater. Sci., Mater. Electron. (USA) vol.12 (2001) p.537–42*]

[69] L. Chen, S. Bagchi, S.J. Krause, P. Roitman [*Proc. IEEE Int. SOI Conf. October 1999, p.123*]

[70] J.S. Jeoung, R. Evans, S. Seraphin [*Mater. Res. Soc. Symp. Proc. (USA) (Spring, 2002)*]

[71] Y. Li, J.A. Kilner, A.K. Robinson, P.L.F. Hemment, C.D. Marsh [*J. Appl. Phys. (USA) vol.70 no.7 (1991) p.3605–12*]

[72] M.J. Current et al. [in *Ion Implantation Science and Technology* Ed. J.F. Ziegler (Ion Implantation Technology Co, Maryland, 2000) p.133–71]; http://www.cityu.edu.hk/ap/plasmausers

[73] A. Ogura [*J. Electrochem. Soc. (USA) vol.145 (1998) p.1735*]

[74] A. Nejim, P.L.F. Hemment [*Crucial Issues in Semiconductor Materials and Processing Technologies* (Kluwer Academic Publishers, 1992) p.225–32]

[75] G. Ryding, T.H. Smick, M. Farley, B.F. Cordts, R.P. Doland, L.P. Allen [*Proc. 11th Conf. Ion Implantation Technology* Austin, Texas, June 1996, p.436–9]

[76] M.J. Anc, J.G. Blake, T. Nakai [*Electrochem. Soc. Proc. (USA) vol.99-3 (1999) p.51–60*]

[77] M. Chen, X. Wang, J. Chen, X.H. Liu, Y. Dong, Y.H. Yu [*J. Mater. Res. (USA) vol.17 no.7 (2002) p.1–10*]

[78] M. Chen et al. [*J. Vac. Sci. Technol. B (USA) vol.19 no.2 (2001) p.337–43*]

[79] A. Ogura, H. Ono [*Appl. Surf. Sci. (Netherlands) vol.150/160 (2000) p.104*]

[80] J. Chen, X. Wang, M. Chen, Z.H. Zheng, Y.H. Yu [*Appl. Phys. Lett. (USA) vol.78 no.1 (2001) p.73–5*]
A. Auberton-Nerve, A. Wittkower, B. Aspar [*Nucl. Instrum. Methods Phys. Res. B (Netherlands) vol.96 (1995) p.420*]

[81] A. Ogura [*Electrochem. Soc. Proc. (USA) vol.99-3 (1999) p.61–6*]

[82] K. Sakaguchi, T. Yonehara [*Solid State Technol. (USA) vol.43 no.6 (2000) p.88–92*]; http://www.canon.com/eltran/tec/tec_com.html

[83] A. Ogura, H. Ono [*Appl. Surf. Sci. (Netherlands) vol.159 (2000) p.104*]

[84] D.K. Sadana, M.I. Current [in *Ion Implantation Science and Technology* Ed. J.F. Ziegler (Ion Implant Technology Co, Maryland, 2000) p.341–7, ISBN 0-9654207-0-1]

[85] S. Nakashima, T. Kalagama, Y. Miyamura, A. Matsuzaki [*Proc. IEEE SOI Conf. (1994) p.71*]

[86] M.J. Anc, W.A. Krull, G. Ryding [*Electrochem. Soc. Proc. (USA) vol.96-3, p.63–73*]

[87] J.B. Liu et al. [*Appl. Phys. Lett. (USA) vol.67 (1995) p.2361*]

[88] P.L.F. Hemment et al. [*Nucl. Instrum. Methods Phys. Res. B (Netherlands) vol.21 (1987) p.129–33*]

[89] M. Chen et al. [*Appl. Phys. Lett. (USA)* vol.80 no.5 (2002) p.880–2]
      N. Sakudo, K. Tokiguichi, H. Koike, I. Kanomata [*Rev. Sci. Instrum. (USA)* vol.48 (1977) p.762–5]

[90] P.L.F. Hemment [personal communication]

[91] U. Bussmann, P.L.F. Hemment, A.K. Robinson, V.V. Starkov [*Nucl. Instrum. Methods Phys. Res. B (Netherlands)* vol.55 (1991) p.856–9]

[92] P.L.F. Hemment et al. [*Nucl. Instrum. Methods Phys. Res. B (Netherlands)* vol.37/38 (1989) p.766]

[93] M. Chen et al. [to be published]

[94] M. Chen [*LII* (2002)]

[95] J. Jablonski et al. [*Electrochem. Soc. Proc. (USA)* vol.96-3 (1996) p.47–9]

[96] J.H. Freeman [*Radiat. Eff. (UK)* vol.100 (1986) p.161–248]

[97] K. Izumi [*Vacuum (UK)* vol.42 no.5/6 (1990) p.333–40]

[98] [*Proc. Conf. Ion Implantation Science and Technology* Trento, Italy, 1978]

[99] D.J. Chivers [*Rev. Sci. Instrum. (USA)* vol.63 no.4 (1992) p.2501]

[100] P.L.F. Hemment, E. Maydell-Ondrusz, K.G. Stephens, J. Butcher, D. Ioannou, J. Alderman [*Nucl. Instrum. Methods (Netherlands)* vol.209/210 (1983) p.157–64]
      J. Blake [in *Encyclopedia of Physical Science and Technology* (Academic Press, 2001) 3rd edition, vol.14]

[101] K.J. Reeson, P.L.F. Hemment [*New Sci. (UK)* November (1987) p.39–43];
      R. Dettmer [*IEE Electron. Power (UK)* April (1987) p.273–7]

[102] P.L.F. Hemment [personal communication]

[103] NASDAQ; C.L. Lin [private communication]

[104] J. Blake [personal communication]

[105] K. Tokiguchi et al. [*Proc. 13th Conf. Ion Implantation Technology* Ed. H. Ryssel, Alpback, Austria, September 2000, p.372–5]

[106] N. Sakudo, H. Koike, K. Tokiguchi, T. Seki [*Nucl. Instrum. Methods Phys. Res. B (Netherlands)* vol.55 (1991) p.300–4]

[107] M.I. Current et al. [in *Ion Implantation Science and Technology* Ed. J.F. Ziegler (Ion Implantation Technology Co, 1996) p.92–174]

[108] A. Chediak, K. Scott, P. Zhang [in *Mater. Sci. Eng. (Switzerland)* vol.225 April (2002)]

[109] See http://www.soitec.com/unibond.htm

Chapter 4

SIMOX/SOI processes: flexibility based on thermodynamic considerations

A. Ogura

4.1 INTRODUCTION

Two major breakthroughs in the history of SIMOX (separation by implanted oxygen) technology led to improved crystalline quality of the top silicon layer. One was the high-temperature annealing (at over 1300°C) after oxygen implantation [1], and the other was the invention of low-dose SIMOX [2]. High-temperature annealing reduced the number of dislocations remaining in the SOI layer of standard dose SIMOX from $10^9/cm^2$ in early material to $10^6/cm^2$, which could be achieved routinely in volume fabrication. Low-dose SIMOX processes further reduced the dislocation density by 3–4 orders of magnitude, and moreover greatly reduced the wafer fabrication cost. The discovery of the internal thermal oxidation (ITOX) process strengthened the integrity of the buried oxide (BOX) film [3] such that the low-dose SIMOX can support mainstream IC technology.

Although fabrication of SOI structures by implanting O^+ ions and annealing at high temperature may appear simple, complex physics aspects need to be taken into account to refine the basic SIMOX process to achieve the desired quality of the layers. In this chapter, detailed analyses of thermodynamic processes in SIMOX are discussed with emphasis on forming high quality SIMOX layers through sophisticated annealing of silicon implanted with oxygen doses lower than the optimum process window. A concept of novel SOI structures emerging from innovative, elaborated annealing schemes is also introduced.

4.2 THERMODYNAMIC PROCESSES IN SIMOX

In traditional standard-dose SIMOX (sometimes called "full-dose" SIMOX), doses of implanted oxygen exceed peak concentrations required for SiO_2 stoichiometry, as shown in FIGURE 4.1 [2]

63

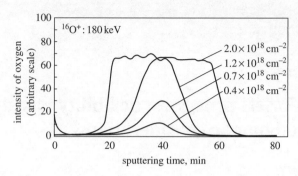

FIGURE 4.1   Oxygen profiles measured by AES after SIMOX implantation with various doses at 180 keV [4].

in the Auger Electron Spectroscopy profiles of oxygen implanted at 180 keV. The formation of SOI structure with stoichiometric doses is essentially completed immediately after implantation. The subsequent high-temperature annealing is performed not to form BOX film but to improve the SOI crystalline quality by recovering implantation damage (Chapters 2 and 3 in this book).

In low-dose SIMOX, the concentration of implanted oxygen atoms is lower than in stoichiometric SiO_2. Implanted oxygen atoms first form SiO_2 precipitates by aggregating at nucleation sites, and then the precipitates grow and coalesce into a continuous BOX layer. Process parameters must be well controlled to form high quality continuous SOI structures with low implanted doses since there is only a narrow process window, called the "dose window", where one can fabricate a continuous BOX at a given energy [2]. For example, the dose window at energy 180–200 keV is $4 \pm 0.5 \times 10^{17}/cm^2$, and with such a narrow range of favourable doses high production yield is not easy to assure.

The low-dose SIMOX process is a thermodynamical process involving oxygen precipitation, growth of precipitates, and coalescence of precipitates into a continuous BOX layer during high-temperature annealing following implantation. The density and arrangement of the nucleation sites for the precipitation play a very important role in the process of formation of low-dose SIMOX. Potential nucleation sites tend to be distributed at the depth of the projection range (R_p) of O^+ implantation since the oxygen concentration is the highest there, or around the damage peak (D_p), where the crystalline damage from the implantation is at a maximum. The calculated depths for R_p and D_p for oxygen implanted at 180 keV are 435 nm and 310 nm, respectively. During high-temperature annealing some precipitates survive and grow larger, while others disappear by transferring oxygen to the growing precipitates. BOX formation is successful when the surviving precipitates coalesce into a continuous film. FIGURE 4.2 illustrates the mechanism

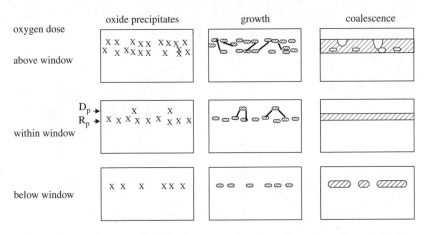

FIGURE 4.2   Oxygen precipitation, growth and coalescence of precipitates during SIMOX annealing with various oxygen doses.

of the precipitation process with the oxygen doses above (top), within (centre) and below (bottom) the dose window, respectively [5]. As shown in this figure, when the oxygen dose is within the dose process window, precipitation leads to the formation of a continuous BOX film. With oxygen doses above the dose window, too many nucleation sites exist around both R_p and D_p, and too many precipitates survive at both. This overabundance of nucleation sites causes simultaneous coalescence of many precipitates at both depths, resulting in trapping of silicon and formation of large Si islands within the BOX film. With oxygen doses below the process window, independent SiO_2 islands are formed instead of a continuous BOX film due to the insufficient concentration of oxygen around precipitation sites.

The precipitation process also affects defect formation in SOI. As shown in FIGURE 4.2, dislocations bridging the precipitates appear in the early stage of annealing. Most of them disappear during the high-temperature annealing; however, those with a very stable configuration remain as the embedded defects in the SOI layer even after lengthy annealing. Reflecting this process, the residual defect density in the low-dose SIMOX is not uniformly distributed across the film but increases when approaching the depth of the BOX. The silicon dislocation profile in low-dose SIMOX is shown in FIGURE 4.3 [6]. It is apparent that the defect density between the D_p location and the BOX film is higher than near the surface of the sample.

The above thermodynamic processes of oxygen precipitation, growth and coalescence of oxide precipitates strongly affect the quality of low-dose SIMOX, i.e. continuity of the BOX and defects in the superficial SOI layer. Using them with consideration is very important not only in designing manufacturable low-dose

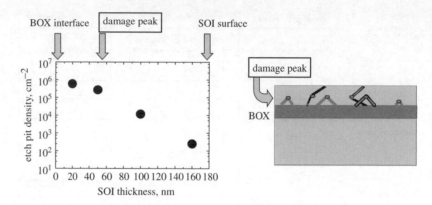

FIGURE 4.3   Defect profile in low-dose/ITOX SIMOX.

processes but also in the development of novel SOI structures and more versatile SOI substrates.

4.3   EXTENSION OF LOW-DOSE PROCESS WINDOW

One of the greatest challenges in fabrication of SIMOX is to lower the implanted oxygen dose to reduce the cost of the substrate, and to extend the dose process window to form high quality SOI layers with good yield. Technological methods stimulating the desired formation mechanisms to improve the continuity of the thin BOX and ensure low SOI defect density will be described in the following sections.

4.3.1   Production worthy methods

Two methods were found particularly effective in improving the integrity of the thin BOX and became part of the commercial SIMOX processes.

First, it was shown that an oxygen rich atmosphere at the high-temperature anneal leads to the internal thermal oxidation (ITOX) of the BOX, and thus greatly improves the continuity and micro-structure of thin BOX SIMOX for doses beyond the optimum range [3,7] (Chapters 2 and 3 in this book). This process was commercialised as ITOX/SIMOX.

Another method of stimulating formation of a continuous BOX with lower than conventional oxygen doses was found to rely on the application of a very low-dose room temperature implant following hot implantation of the low-dose. By amorphising the region near Rp by implanting a dose of around 1×10^{15}/cm^2 at room temperature and epitaxial regrowth during high-temperature annealing, the continuous BOX can be formed with the doses lower than the optimum process window [8]. The amorphisation step provides the template for the growth of the BOX, and with addition of ITOX allows greater flexibility in achieving final BOX thickness. This process has been commercialised as modified low-dose (MLD) SIMOX (Chapter 6 of this book).

4.3.2 Dose-window extension by controlling precipitation

The above two commercial methods extensively employ annealing in highly oxidising ambient. In an extreme case the continuous BOX may be formed with very low implanted doses by annealing in low oxygen concentration ambient and controlling the precipitation process with the temperature ramp rate. The following considerations show such a possibility by implementing careful analyses of the precipitation process.

It is known that the minimum size of stable precipitate increases with temperature [9], and also the rate of temperature change affects growth of precipitates, as illustrated in FIGURE 4.4 [5]. At lower ramp rates small precipitates survive due to the higher chance of growing to the critical size at the next slightly higher temperature, while at high ramping rates small precipitates are eliminated by dissolution, and the oxygen that they release contributes to the growth of larger precipitates. Thus, by changing temperature ramping rate in the annealing process the thermodynamic processes of growth of oxide precipitates, i.e. the BOX film formation, can be controlled.

The impact of temperature ramp rate on the extension of the dose process window at 180 keV is shown in FIGURE 4.5 [10].

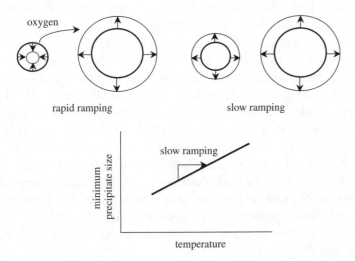

FIGURE 4.4   Control of precipitation growth by changing temperature ramping rate during annealing.

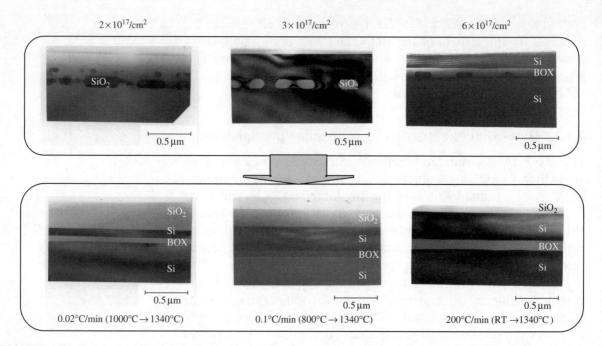

FIGURE 4.5   Summary of dose-window extension by controlling ramping rate during high-temperature annealing.

Under conventional annealing conditions in an atmosphere of $Ar/O_2 = 100/0.5$ and with the typical furnace ramp rates of 20°C/min a discontinuous BOX forms with the implanted doses $2 \times 10^{17} O^+/cm^2$ and $3 \times 10^{17} O^+/cm^2$. These doses are lower than the optimum process window for the energy of 180 keV. For such low-doses annealing with lower ramping rates encourages growth of relatively few incipient oxide precipitates that may finally coalesce to form a continuous buried oxide. In the case of the $3 \times 10^{17} O^+/cm^2$ dose, the ramping rate in the temperature range of 800°C to 1340°C was 0.1°C/min, while other conditions were the same as in the standard annealing process. As a result, a complete SOI structure with a continuous BOX film, similar to the one obtained with a dose within the dose-window conditions, was fabricated. In the case of the $2 \times 10^{17} O^+/cm^2$ sample, annealing with a very low ramping rate of 0.02°C/min from 1000°C to 1340°C had to be applied to produce an SOI structure with a continuous BOX film. In both cases low ramping rate increased the number of oxygen nucleation sites that survived the temperature changes and contributed to the coalescence of buried oxide. This approach resulted in extension of the dose process window toward lower doses.

In contrast, for the doses above the process window, high ramping rates promote dissolution of precipitates smaller than the critical size and reduce the overall number of precipitates, which in turn results in formation of the BOX film without Si islands. Indeed, a rapid increase of temperature at a 200°C/min rate applied to the sample implanted with the dose of 6×10^{17} O$^+$/cm^2 resulted in formation of a continuous SOI structure, whereas annealing with a standard ramp rate of 20°C/min would result in Si island rich buried oxide, as shown in FIGURE 4.5.

4.4 USING ATMOSPHERIC OXYGEN IN BOX FORMATION

As in ITOX, which is actually high-temperature oxidation, diffusion of atmospheric oxygen in silicon can be used to enhance precipitation and growth of precipitates. This can be achieved by controlling the atmospheric oxygen concentration during high-temperature annealing in conjunction with controlling the ramping rate [11]. Since both in- and out-diffusion of oxygen occurs during high-temperature annealing in SIMOX [12], it is reasonable to expect the oxygen atoms introduced from the atmosphere through the surface to contribute to oxygen precipitation and growth of the precipitates. The growth of oxide precipitates in Si is described by EQN (4.1) [13]:

$$V(T, t) = \frac{8\pi \sqrt{2}}{3} \left[\frac{C_I - C_E}{C_S - C_E} D(T)t \right]^{2/3} \qquad (4.1)$$

where $V(T, t)$ is the precipitate volume, T is the annealing temperature, t is time, C_I and C_E are the initial and equilibrium concentrations of oxygen interstitials, respectively, C_S is the atomic concentration of oxygen in the precipitates, and $D(T)$ is the oxygen diffusion coefficient. Oxygen introduced from the atmosphere should increase the C_I and therefore enhance the growth of precipitates as well as increase the temperature (T) or time (t), which corresponds to low ramping rate.

The effect of atmospheric oxygen on BOX film formation is demonstrated in FIGURE 4.6 [11]. When a slightly higher ramping rate than that in FIGURE 4.6, 0.03°C/min from 1000°C to 1340°C, was used to anneal the sample implanted with an oxygen dose of 2×10^{17}/cm^2 in ambient with Ar/O$_2$ ratio of 100/0.5, independent SiO$_2$ islands were observed instead of buried oxide film. When the oxygen concentration in the atmosphere was increased, these oxide islands became larger, and a continuous BOX film eventually appeared with an Ar/O$_2$ ratio of 100/1. This shows that the precipitation process can be controlled by changing oxygen

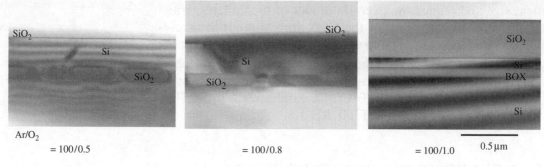

FIGURE 4.6   Effect of oxygen concentration in atmosphere on BOX formation.

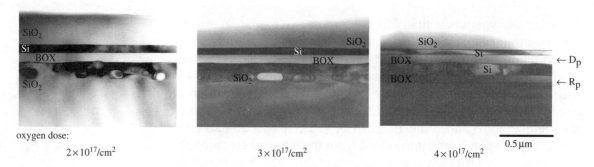

FIGURE 4.7   Novel structures with BOX film at D_p and double BOX layers.

concentration in the atmosphere as well as by changing the ramping rate in the annealing.

Increasing the ramping rate and concentration of oxygen in the annealing ambient even further resulted in formation of a completely different structure, as shown in FIGURE 4.7 [11]. In this structure, a BOX film formed at a depth of D_p instead of R_p, and oxide islands emerged at a depth of R_p below the continuous BOX. Dislocations bridging the BOX layer and islands at R_p were observed, as in the case of conventional SIMOX. In this annealing scheme, sufficiently high oxygen concentration in the ambient promoted the diffusion of oxygen into the silicon and growth of precipitates at the damage peak, since this is closer to the surface than the peak of concentration of implanted oxygen. A low ramping rate, which also enhances precipitate growth, is essential for forming the BOX layer at D_p. In this SOI structure defects and oxide islands underneath the BOX layer should not degrade device characteristics since these are fabricated in the superficial Si layer. Moreover they may act as efficient gettering sites for metallic impurities introduced during processing. This structure may also provide a lower defect density SOI substrate compared to a conventional low-dose SIMOX structure, since the most defective area is replaced by the BOX layer [6].

An interesting SIMOX structure consisting of double BOX layers with continuous BOX at both D_p and R_p, can be fabricated by one implantation of the dose from the process window (4×10^{17} O^+/cm^2 at 180 keV) and annealing using a low ramping rate and high oxygen concentration in the ambient. After annealing, such a structure consists of surface SiO_2, a superficial Si layer, a first BOX layer, a sandwiched Si layer with a high dislocation density, a second BOX layer, and a Si substrate. Three SOI structures implanted with oxygen doses of 2×10^{17}, 3×10^{17} and $4 \times 10^{17}/cm^2$ and annealed with the same ramping rate and the same Ar/O_2 ratio are shown in FIGURE 4.7. As shown in this figure, continuous BOX layers formed at D_p at all doses but increasing the dose of implanted oxygen increased the volume of the SiO_2 islands at R_p. Surprisingly, the thicknesses of the superficial Si and BOX layers were found to be almost the same at all doses. This suggests that the majority of oxygen atoms constituting the BOX layer at D_p originated from the atmosphere rather than from the implantation, while implanted oxygen contributed to the BOX layer at R_p.

To verify the origin of oxygen in the BOX film, the $^{16}O/^{18}O$ profiles in the SOI structure fabricated by the conventional low-dose SIMOX process without ITOX were measured by secondary ion mass spectroscopy (SIMS). The results are shown in FIGURE 4.8(a). The natural abundance of ^{18}O is about 0.2%. In the SIMOX process, only ^{16}O is implanted into Si by using mass separation. Therefore, if only implanted oxygen contributed to the formation of the BOX the film should consist exclusively of ^{16}O. However, there was some ^{18}O found in the film, even in the conventional low-dose SIMOX. The $^{16}O/^{18}O$ ratio was almost the natural abundance at the surface of the surface oxide, and the ^{18}O concentration decreased further into the depth of the substrate.

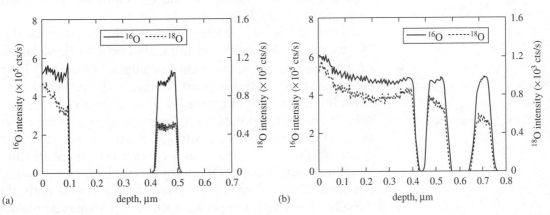

FIGURE 4.8 $^{18}O/^{16}O$ profiles in (a) SOI structure fabricated by conventional low-dose SIMOX without ITOX, and (b) SOI with double BOX layers.

This finding leads to the conclusion that exchange of implanted oxygen with atmospheric oxygen occurs even in the conventional low-dose SIMOX process. The ^{18}O ratio in the BOX film formed at D_p was higher compared to the conventional low-dose SIMOX and close to the ratio of natural abundance. In the double BOX structure, the ^{18}O ratio in the BOX film at R_p was almost the same as that in the BOX film with conventional low-dose SIMOX, but the ^{18}O profile in the BOX film at D_p was similar to that of the surface oxide of the conventional case (FIGURE 4.8(b)). With an oxygen dose of $2 \times 10^{17}/cm^2$, the ^{18}O ratio in the BOX film at D_p was almost the same as that in the surface oxide. This also supports the idea that the oxygen in the BOX film at D_p originates from the atmosphere. These results indicate that the BOX film at D_p in these new structures mainly consisted of oxygen from the atmosphere, while the BOX film at R_p consisted of implanted oxygen, with some exchanges with atmospheric oxygen.

These studies show that it may be possible to form an SOI structure by simply introducing damage into the substrate and following with annealing in oxygen containing ambient in order to induce precipitation at defect range.

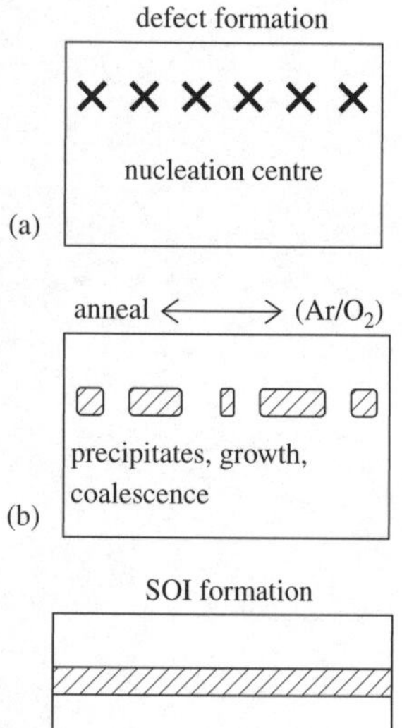

4.5 EMERGING NOVEL SOI STRUCTURES

The concept of a novel SOI fabrication process, which actually may no longer be SIMOX, is illustrated in FIGURE 4.9 [14]. Crystalline defects are introduced at a controlled depth in the Si substrate to create nucleation centres for oxygen precipitation. The sample is then annealed in an Ar/O_2 atmosphere to enhance oxygen precipitation, that with a sufficient supply of oxygen would grow and coalesce during annealing to complete the formation of a continuous BOX layer.

To realise the idea, implantation of light ions such as He^+ and H^+ was used. Although the implanted dose of light ions is still of the order of $10^{17}/cm^2$, light ion implantation (LII) is attractive because of smaller implant damage and a potential for easier annealing even after implantation at room temperature. Not only does the room-temperature implantation appear to be less costly, but also the implantation profile is sharp and easy to control by changing the acceleration energy, such that more flexible SOI structures and thinner SOI and/or BOX films, can eventually be obtained. The implanted light ions diffuse out in the early stage of the subsequent high-temperature annealing, so they do not degrade the performance of devices fabricated in SOI. The LII can also be combined with conventional SIMOX with O^+ implantation, so

FIGURE 4.9 Concept of eliminating O^+ implantation from SIMOX.

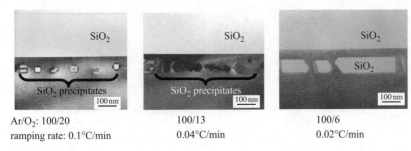

Ar/O$_2$: 100/20   100/13   100/6

ramping rate: 0.1°C/min   0.04°C/min   0.02°C/min

FIGURE 4.10   Oxygen precipitation at He$^+$ implantation damage sites and growth of precipitates by annealing in oxygen including atmosphere.

it may be possible to realise innovative processes and structures with smaller O$^+$ doses and/or thinner BOX films and patterns. FIGURE 4.10 shows that the implantation damage induced by He$^+$ implantation can act as nucleation centres for the oxygen precipitation. In the experiment, the acceleration voltage was set to 45 keV so that the R$_p$ was almost the same as for O$^+$ implantation at 180 keV. The structures strongly depended on the annealing conditions, particularly the ramping rate and the oxygen concentration in the atmosphere. The pictures shown in FIGURE 4.10 are the TEM images of samples with a He$^+$ dose of 1×10^{17}/cm^2 after annealing at a ramping rate of 0.1°C/min, 0.04°C/min and 0.02°C/min, respectively. The Ar/O$_2$ ratios were also varied as 100/20, 100/13 and 100/6, so that the surface oxides have almost the same thickness. Oxygen precipitation apparently occurred at the damage induced by the He$^+$ implantation, and the precipitates become larger as the ramping rate was reduced. The precipitates probably grew and coalesced during the low-ramping-rate annealing. However, it can be noted that further reduction in the ramping rate for a sample with the same implantation dose did not produce a continuous BOX film. This implies that the density of nucleation centres was not sufficient to form a continuous BOX film. Increasing the dose of He$^+$ increased the volume of precipitates, and a continuous BOX layer was finally formed at a dose of 4×10^{17}/cm^2 with the same ramping rate and oxygen concentration, as shown in FIGURE 4.11.

To evaluate the SOI formation mechanism in this process, the early stage of the annealing was observed by TEM. Micro cavities were formed at the depth of implantation damage prior to oxygen precipitation. The oxygen in the annealing atmosphere should have diffused into the substrates during the high-temperature annealing and might have reacted with the dangling bonds at the inner surface of the cavities. Once the SiO$_2$ film formed at the surface, it is reasonable to believe that the Si/SiO$_2$ interface acted as a site for efficient reaction with diffused oxygen, as occurs with ITOX. This

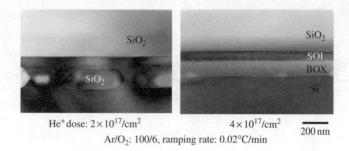

He$^+$dose: $2 \times 10^{17}/\text{cm}^2$ $4 \times 10^{17}/\text{cm}^2$ 200 nm

Ar/O$_2$: 100/6, ramping rate: 0.02°C/min

FIGURE 4.11   SOI structure fabricated by He$^+$ implantation and annealing.

supports the formation mechanism of the SOI structure with the process demonstrated in this study.

4.6   PARTIAL SOI FORMATION

A partial SOI substrate lends itself to the system-on-chip (SOC) technology because both high-speed logic circuits in the SOI region and high-performance DRAM cells in the bulk region can be built simultaneously [15]. Moreover, patterned SOI may stimulate device design with selective use of SOI regions to reduce junction capacitance without the problems of the floating body effect and to overcome self-heating effects [16].

At present the most promising technique for fabricating partial SOI is perceived to be patterned SIMOX in which O$^+$ ions are implanted through a patterned mask [17]. A major challenge for this technique relies in reduction of defect density at the boundaries between the SOI and non-SOI regions. Techniques used to improve the quality of SIMOX are applicable to partial SOI formation. Lower defect density and smaller defective area at the mask edge can be achieved with low-dose SIMOX compared to the standard-dose SIMOX. Both two-step implantation (hot and RT) and ITOX are effective in improving the quality of partial SOI, especially in that they can be applied selectively in the desired area of the wafer (Chapter 6 in this book). In addition, partial SOI can be formed with selective use of controlled precipitation induced by implantation of light ions. Examples will be shown in the following section.

In the experiment, He$^+$ was implanted through a 0.5-μm-thick patterned SiO$_2$ mask, with a 45 keV acceleration voltage and $(3.5–4) \times 10^{17}/\text{cm}^2$ doses. A typical partial SOI structure fabricated using the technique is shown in FIGURE 4.12 [18]. A BOX layer is formed in the Si substrate. Although the BOX layer is thicker at the pattern edge than elsewhere so that the SOI surface is slightly elevated there, the observed density of crystal defects is much less than that formed by conventional patterned SIMOX.

(a)   500 nm

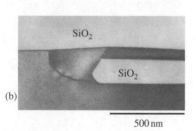

(b)

500 nm

FIGURE 4.12   Partial SOI structure fabricated by LII technique.

With the LII technique, the volume ratio filled by SiO_2 in the BOX film can be changed depending on the annealing conditions. For partial SOI application, the buried insulator does not necessarily need to be completely filled with SiO_2, because it is supported by the surrounding Si. For example, an SOI structure with an ultimately small SiO_2 volume is shown in FIGURE 4.13. This kind of structure is called SON (silicon-on-nothing) rather than SOI. SON structure is usually fabricated using high-aspect trench formation and high-temperature annealing in H_2 ambient [19], or using preferential etching of a sacrificial layer [20]. The advantages of the LII technique lie in its simple fabrication process and SiO_2 coating of the inner surface of the buried empty space. Since the LII technique causes less damage to the Si crystal, potentially fewer defects may develop in SOI compared with conventional SIMOX. Further reduction in defect density may be expected because stress introduced by volume expansion of SiO_2 formation is negligible. As shown in FIGURE 4.13, no defects were observed, even at the pattern edge. The surface of the fabricated SON structure was extremely smooth compared with that of SOI and SON fabricated using other techniques. This smooth surface may also be considered as a great advantage for ULSI-SOC, because state-of-the-art photolithography requires extremely smooth surfaces to avoid impact on focal depth.

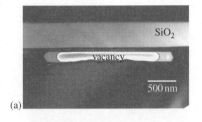

(a)

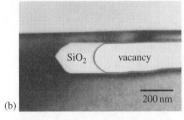

(b)

FIGURE 4.13   Partial SON structure fabricated by LII technique.

4.7   CONCLUSION

In this chapter, the low-dose SIMOX process has been reviewed from the viewpoint of thermodynamics. Various innovative techniques of refinement of the SIMOX process have been introduced and explained based on thermodynamic considerations for generation of nucleation centres, and growth and coalescence of precipitates during the annealing process. The author hopes that readers will have the opportunity to recognise the potential for further developments and refinements of the processes, and this chapter will stimulate further discussions among the scientists.

REFERENCES

[1] G.K. Celler, P.L.F. Hemment, K.W. West, J.M. Gibson [*Appl. Phys. Lett. (USA)* vol.48 (1986) p.532]
[2] S. Nakashima, K. Izumi [*J. Mater. Res. (USA)* vol.8 (1993) p.523]
[3] S. Nakashima et al. [*J. Electrochem. Soc. (USA)* vol.143 (1996) p.244]
[4] By courtesy of Dr. S. Nakashima
[5] A. Ogura [*J. Electrochem. Soc. (USA)* vol.145 (1998) p.1735]

[6] A. Ogura, T. Tatsumi, T. Hamajima, H. Kikuchi [*Appl. Phys. Lett. (USA)* vol.69 (1996) p.1367]

[7] J. Jablonski, M. Saito, Y. Miyamura, T. Katayama [*Proc. Eighth Int. Symp. on SOI Technology and Devices* Ed. S. Cristoloveanu, vol.97-23 (The Electrochem. Soc. Series, Pennington, 1997) p.51–6]

[8] O.W. Holland, D. Fathy, D.K. Sadana [*Appl. Phys. Lett. (USA)* vol.69, (1996) p.674].

[9] R. Johnes (Ed.) [*Early Stage of Oxygen Precipitation in Silicon* (Kluwer Academic Publisher, Dordrecht, The Netherlands, 1996)]

[10] A. Ogura, H. Ono [*Appl. Surf. Sci. (Netherlands)* vol.159–160 (2000) p.104]

[11] A. Ogura [*Appl. Phys. Lett. (USA)* vol.74 (1999) p.2188]

[12] H. Ono, A. Ogura [*J. Appl. Phys. (USA)* vol.87 (2000) p. 7782]

[13] K. Wada, H. Nakanishi, H. Takaoka, N. Inoue [*J. Cryst. Growth. (Netherlands)* vol.57 (1982) p.535]

[14] A. Ogura [*Jpn. J. Appl. Phys. (Japan)* vol.40 (2001) p.L1075]

[15] R. Hannon et al. [*VLSI Tech. Dig.* vol.66 (2000)]

[16] W.S. Chen, L.L. Tian, Z.J. Li [*IEEE Int. Solid-State and Integrated Circuit Technology Conf. Proc.* vol.49 (1998)]

[17] G.M. Cohen, D.K. Sadana [*Mater. Res. Soc. Symp. Proc. (USA)* vol.686 (2002) p.A2.4.1]

[18] A. Ogura [*2002 Int. Conf. on Solid State Devices and Materials* Nagoya, 2002]

[19] T. Sato et al. [*VLSI Tech. Dig.* vol.206 (1998)]

[20] M. Jurczak et al. [*IEEE Trans. Electron Devices (USA)* vol.47 (2000) p.2179]

Chapter 5

Electrical and optical characterisation of SIMOX substrates

H. Hovel

5.1 INTRODUCTION

Silicon-on-insulator has gone far in substantiating its early promise in semiconductor technology. Mainstream computer and data processing products are available, and many more are in development [1]. SOI material continues to evolve to meet the needs of each new technology generation. Two types of starting substrate material have emerged from a larger list of early candidates: SIMOX and bonded SOI, and both continue to show great promise. Of the two, SIMOX manufacturing technology exhibits greater flexibility. The buried oxide (BOX) and superficial Si layer thicknesses are easily modified by changes in the implantation energy, the dose of implanted oxygen atoms, and the annealing atmosphere and time schedule. (Details of SIMOX development and manufacturing are discussed in Chapter 6 of this volume by D. Sadana.) However, unlike "bulk" silicon starting material, in which wafer to wafer, ingot to ingot, and manufacturer to manufacturer consistency and uniformity have been established by many years of development, SOI material still exhibits a great deal of variability.

This evolutionary nature of SOI development has both advantages and disadvantages. On one hand, the material can be tailored to fit the needs and requirements of each new technology node. The SIA Roadmap for Semiconductors [2] describes in detail how the Si and BOX thicknesses, surface roughness, background metal content, and many other material properties must be changed (improved) if the advances in performance, power requirements, device density etc. expected for future technology generations are to be realised. The flexibility of SOI material manufacturing is ideally suited to meet these challenges. An example of this is the switch from "standard dose" material, with implant doses of 1.8×10^{18} per cm^2 and BOX thicknesses around 380 nm, to "low-dose" made with implant doses $\leq 5 \times 10^{17}$ per cm^2 and BOX thicknesses ≤ 150 nm. Future technologies are likely to make use of material with BOX thicknesses less than 100 nm and Si thicknesses less than 30 nm [2].

On the other hand, this means that each new SOI material generation is a relative unknown. As the implant and anneal conditions are modified to create new thickness regimes, new problems can be introduced and old ones exacerbated. In addition, SOI material manufacture is still a batch process. Implants are carried out in a 13 wafer to 25 wafer lot size. Anneals are carried out in 50 to 100 wafer lots. Since problems can occur during both implants and anneals, quality assurance requires that at least one wafer out of every 4 to 5 lots be fully characterised, and more if some problem becomes evident from characterising either this sample wafer or an accompanying monitor bulk wafer. This characterisation is usually time consuming and, of course, removes the wafer from potential use in product applications. However, starting material characterisation is essential for quality control just as it was in the early days of bulk Si wafer development. If problems are discovered, the wafer sampling size (more wafers per lot) must be increased until the problem is eliminated. Characterisation of the physical and electrical properties of SOI material can be very significant in understanding the yield and reliability of the material [3].

Characterisation of starting SOI material takes many different forms. Physical characterisation involves the detection and measurement of defects in the Si and BOX layers. In the oxide, "pinholes" were an early problem in SIMOX and "voids" in bonded material. Pinholes in SIMOX were caused largely by particles which masked small areas from the implanted oxygen and resulted in Si-rich conducting paths through the oxide. These often showed up as short circuit paths through the BOX. The BOX also contained Si-rich precipitates, located near the bottom in standard dose SIMOX and near the centre in low-dose material, as seen in TEM (transmission electron micrograph) pictures. The Si layer contained high densities of threading dislocations, densities of 10^6 to 10^7 per cm^2 for standard dose material and 10^2 to 10^4 for low-dose. The dislocations are studied through SECCO etching. Another defect type unique to SOI discovered in the early days became known as "HF defects," caused mostly by particles which form silicides during the anneal and subsequently cause holes in the Si after HF etching. Pinholes, dislocations and HF defects have all been improved but not eliminated. Detection and metrology of such defects will be discussed in Chapter 6. Some of the electrical consequences will be discussed here.

This chapter will focus mainly on the electrical and optical characterisation of SIMOX starting material for CMOS applications. Optical characterisation is used for determining the thicknesses of the Si and BOX layers as well as their thickness uniformities. Electrical characterisation includes determining the I–V behaviour of the BOX (pinhole densities, leakage currents at high fields and breakdown voltages) and the key electrical parameters of the Si film such as mobilities, charge densities and lifetimes. Different characterisation techniques will be described, especially those used most often by the author for quality control of SOI material used at IBM. Two excellent treatises on characterisation of semiconductor material in general and SOI in particular are available and are recommended to the reader for further detail [4, 5].

5.2 OPTICAL CHARACTERISATION

The most fundamental properties of SOI substrates are the thicknesses of the Si and BOX layers and the uniformity, particularly of the Si layer. All the device and circuit designs take these thicknesses into account, in the junction depths, for example. The threshold voltages depend strongly on the Si thickness for fully depleted devices, and the threshold voltage uniformity therefore depends on the Si uniformity. Short channel effects are also dependent on these thicknesses, and a great deal of time in the SIA Roadmap formulations has been spent on predicting the required thicknesses and uniformities for each technology node. Future technologies may also make use of strained Si/SiGe structures on SOI, as pointed out in the Roadmap. Strained Si material is of interest because of the promise of enhanced performance [6]. In these cases the composition of the SiGe and its uniformity are added considerations, since they affect the degree of strain in the Si layer.

Electrical thickness measurements of the layers are possible but tedious and have higher potential errors than optical methods. Capacitance measurements of the BOX thickness can be made with evaporated metal electrodes on the BOX, for example. These measurements are useful in obtaining the buried oxide effective charge, as discussed in Section 5.3, but the potential error in the thickness measurement is determined by the accuracy of the device area measurement, which is several percent or higher. Electrical measurements of the Si layer thickness are difficult because the overall capacitance of the SOI device is a series combination of three capacitors (metal electrode to Si, Si to BOX, BOX to substrate).

Optical measurements are an elegant and highly accurate method for determining the layer thicknesses and uniformities with accuracies of better than 0.5%. Silicon and SiGe layers are semitransparent in the visible and near-infrared wavelength regions, as is the BOX. FIGURE 5.1 shows a typical SOI structure

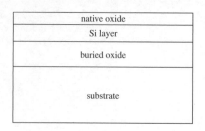

FIGURE 5.1   Typical SOI material stack.

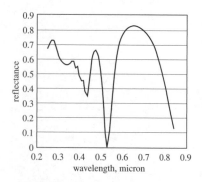

FIGURE 5.2   Reflectance at normal incidence of a typical SOI material stack.

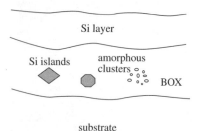

FIGURE 5.3   Real SOI structure with wavy surfaces and interfaces (exaggerated), Si islands and amorphous Si clusters in the BOX.

consisting of a native oxide on the surface, the Si film residing on a buried oxide layer, and the underlying Si substrate. In strained Si structures on SOI, one or more SiGe regions may also be present between the BOX and the Si film. The interfaces between these materials involve abrupt changes in the refractive indices resulting in optical interference patterns with maxima and minima and peak widths used to determine thicknesses and compositions. FIGURE 5.2 shows a typical reflection spectrum for a SIMOX substrate with a 1.2 nm native oxide, a 120 nm Si layer, and a 140 nm BOX. The rich amount of detail allows for accurate and reproducible thickness determination. Spectra like these are analysed by matching the measured spectrum to ones calculated using the Fresnel coefficients for material stacks like those shown in FIGURE 5.1. Regression iterations are carried out until the calculated and measured spectra match as closely as possible, and the final thicknesses are those needed to produce the best goodness of fit.

A number of manufacturers have produced instruments which measure the reflection spectrum and determine the thicknesses automatically. They all allow for multiple measurement points across a wafer, resulting in statistics on layer uniformity and in some cases in a wafer thickness map with a resolution depending on the number of measurement points. One manufacturer uses special optics and software to measure up to 30,000 points simultaneously, allowing highly detailed thickness maps of the Si layer to be obtained at a small cost in accuracy.

As the thicknesses of the layers decrease, pure reflection measurements begin to become less accurate because of the complex nature of the BOX layer and the various interfaces. FIGURE 5.3 shows a schematic of an SOI stack in real SIMOX material. The added features that need to be accounted for are (1) a wavy interface between the Si and BOX, (2) a wavy interface between the BOX and substrate, (3) Si precipitates of a range of sizes within the BOX layer, and (4) areas of Si richness (and/or oxygen deficiency) sometimes characterised as amorphous Si clusters. Spectroscopic ellipsometry (SE) is suitable for analysing such a complex structure accurately. The wavy interfaces are modelled as extra layers with compositions between Si and SiO_2, while the BOX is modelled as a sum of 3 different SiO_2 layers, each with different Si content. An accurate model for low-dose SIMOX consists of at least 6 different layers. Standard dose SIMOX requires a 5-layer BOX but the interface roughness (waviness) is less significant and is usually ignored. Spectroscopic ellipsometry can account for these complexities because of the larger amount of available measurement information and the very good signal to noise capability.

In spectroscopic ellipsometry, the intensities of reflected light polarised perpendicularly (r_s) and parallel (r_p) to the plane of incidence are measured. The ratio of these two reflectances is used to obtain the ellipsometric parameters Δ (del) and ψ (psi), where

$$\frac{r_p}{r_s} = \tan \psi \exp(i\Delta) \qquad (5.1)$$

To obtain the thicknesses, a model is constructed (a material stack such as FIGURE 5.1) and the resulting calculated optical spectra are compared to the data, iterating as many times as necessary to obtain the best goodness of fit.

FIGURE 5.4 shows an ellipsometry spectrum, plotted as $\tan \psi$ and $\cos \Delta$, and FIGURE 5.5 shows the SOI structure resulting from the analysis. The top layer called the native oxide accounts for both the surface SiO_2 and about 0.6 nm of surface roughness. The 122 nm Si layer is followed by a 0.3 nm interface waviness, and the BOX is divided into regions A, B, and C to account for the different excess Si content. Such a detailed model yields a nearly perfect match of simulated and experimental data. Both the interface roughness and a 3-layer BOX are necessary for accurate measurements. If a single layer BOX is assumed, the calculated BOX thickness can be in error by as much as 5% to 6%.

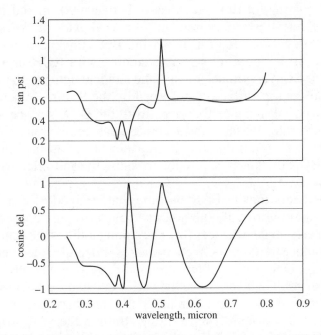

FIGURE 5.4   Measured spectroscopic ellipsometry spectra for SIMOX with 120 nm Si on 134 nm BOX.

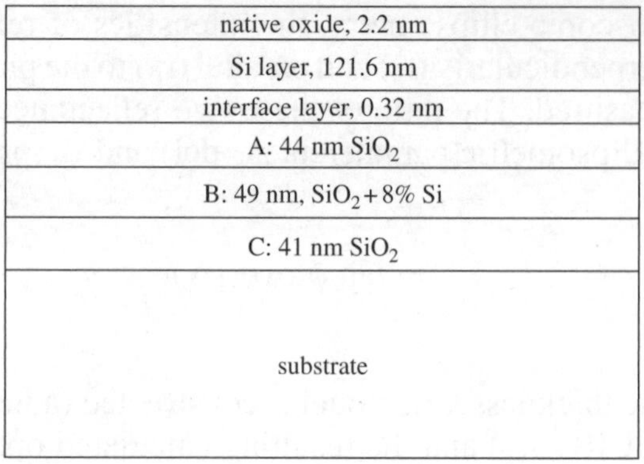

native oxide, 2.2 nm
Si layer, 121.6 nm
interface layer, 0.32 nm
A: 44 nm SiO_2
B: 49 nm, SiO_2 + 8% Si
C: 41 nm SiO_2
substrate

FIGURE 5.5   SIMOX structure obtained from the analysis of the spectra in FIGURE 5.4.

Even the model of FIGURE 5.5 with its 3-region BOX can lead to errors in the BOX thickness under certain conditions. The apparent total BOX thickness measured in the complete structure with the Si layer intact is slightly different if measured after the Si layer is removed. This discrepancy has been noted for both ellipsometry and reflectometry measurements, and for both bonded and SIMOX wafers. It may be due to the very low amount of light deriving from the BOX/substrate interface, since the refractive indices of SiO_2 and Si are such that the BOX forms a very good antireflective coating for BOX thicknesses between 90 nm and 130 nm. FIGURE 5.6 shows measurements of buried oxide thicknesses for 18 SIMOX lots (sorted for increasing BOX thickness) measured with the Si on and after selective etching with KOH or TMAH. When the Si layer is intact, most of the incident light is either reflected from the top surface or absorbed by the Si film. When the Si is removed, all the signal derives from the BOX surface and the BOX/substrate interface, and accurate measurements are obtained. FIGURE 5.6 shows that the discrepancy seems to be reduced at higher BOX thicknesses, possibly because the anitreflective effect is not as strong.

It has been known for some time that strained Si can lead to enhanced mobilities of both holes and electrons and consequently higher performance devices [6]. The strain modifies the band structure to reduce the carrier effective masses and reduce the intervalley scattering. One way to obtain strained Si is to epitaxially grow a Si film on a SiGe layer. Since the lattice constant of SiGe is larger than that of Si, the Si atoms in the Si film are displaced from their equilibrium positions. The resulting strain in the Si is a function of the Ge composition in the SiGe film and the degree to which the lattice of the SiGe has "relaxed" to its equilibrium structure,

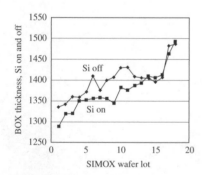

FIGURE 5.6   Comparison of BOX thicknesses measured with the Si layer intact and with the layer removed. The wafer lots were sorted for increasing BOX thickness.

as opposed to being strained itself by the underlying substrate. Strained Si devices can be produced on either bulk or SOI material, but layers grown on SOI have the advantages of SOI (isolation, reduced junction capacitances, reduced short channel effects, etc.) in addition to the enhanced mobility.

Spectroscopic ellipsometry is ideally suited for measuring the thicknesses in strained Si material, either on bulk substrates or SOI. The SiGe layers may have contents ranging from 10% to 35% Ge with thicknesses of tens to hundreds of nanometres. The presence of the Ge alters the optical properties of Si substantially in the UV and visible wavelength regions, facilitating the measurement and analysis of Si/SiGe/BOX/substrate material stacks and similar structures. Another technique that has been very successful for examining strained Si structures is glancing X-ray reflection (GXR), in which short wavelength X-rays (usually 1.54 angstrom) are incident on the wafer and scanned at angles of 0 to 3.5°. Interference patterns are obtained as in optical reflection, but in addition, the spectra are very dependent on the surface and interface roughnesses. GXR has the advantage over SE that the roughnesses are obtained directly instead of being modelled indirectly as additional thin layers in the ellipsometry regressions. However, GXR measurements can require more measurement time per wafer position than ellipsometry, and GXR measurements are restricted to film stacks with total thicknesses less than several hundred nanometres. GXR measurements are also not accurate for determining BOX thicknesses because the optical constants of Si and SiO$_2$ are very close at X-ray wavelengths. Faster GXR measurement tools have been reported recently. Both GXR and SE metrology using multi-channel analysers can reduce the measurement time per spot to less than one minute.

FIGURES 5.7 and 5.8 show GXR and SE measurements on the same strained Si/SiGe substrate. The X-ray measurement indicates a surface roughness of 1.5 angstrom at the top surface and 1.1 angstrom at the Si/SiGe interface. Both the SE and GXR can determine the Ge composition as well as the layer thicknesses. The composition is determined in SE by comparing the measured spectra with known spectra of SiGe layers with various compositions, and is determined in GXR by the densities of the measured films. Both types of measurement give compositions which are within a few percent of the compositions determined by X-ray diffraction (XRD). The strain and degree of relaxation of the SiGe film can be determined by Raman spectroscopy [7] coupled with an independent measure of the Ge composition by SE, GXR, SIMS or XRD. Raman measurements can also detect the degree of strain in the Si layer on SiGe [8].

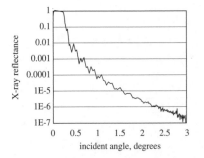

FIGURE 5.7   Glancing X-ray reflectance versus incidence angle for a strained Si layer on SiGe on SOI.

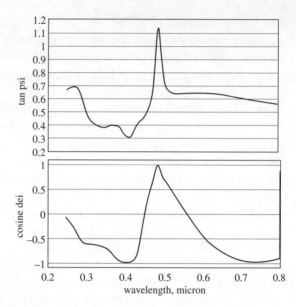

FIGURE 5.8   Spectroscopic ellipsometry spectra for the sample of strained Si on SiGe on SOI.

The thickness uniformity is an important quality control measure of the material. The threshold voltage and other electrical parameters can depend on the Si thickness, especially in fully depleted devices. In strained Si material, the degree of strain in the film, and therefore the mobility, may depend on the Si and SiGe thicknesses and the uniformity of the SiGe composition. Spectroscopic reflectance or ellipsometry can be carried out at various positions on SOI wafers to obtain low resolution thickness maps, and SE can yield the thicknesses and compositions at each position. Typically, measurements are made at 49 or more points across a 200 mm wafer and statistics such as average thickness, standard deviations, and max to min thicknesses are obtained.

For higher resolution maps, the Acumap tool (ADE Corporation) can produce maps with 30,000 points across a 200 mm wafer. This tool uses special optics to illuminate the wafer at a large number of points simultaneously while measuring the reflectance at 25 wavelengths from the visible to near-IR using narrow bandpass filters. The measured spectra at these 25 wavelengths must be matched to spectra libraries created for SOI film stack models. Thickness mapping of one or more layers has proven to be valuable in finding and diagnosing problems with the implant, anneal, or both, and maps of the Si layer thickness have been useful in developing optimised implanter designs and operation. Problems such as striping, nonuniform implant dosing, improper wafer mounting, and others have been discovered this way. Edge exclusion effects can also be studied this way.

TABLE 5.1   Si layer thickness standard deviations versus edge exclusion,
low-dose SIMOX.

Edge exclusion, mm	Standard deviation, nm	Edge exclusion, mm	Standard deviation, nm
0	1.13	4	1.0
1	1.17	5	0.98
2	1.15	6	0.95
3	1.15	8	0.92

TABLE 5.1 shows the Si thickness standard deviations for low-dose SIMOX material as a function of edge exclusion, and demonstrates the high degree of uniformity obtainable. Typical max to min Si thicknesses across 200 mm SIMOX wafers are 3–4 nm for 70–120 nm Si films. Tighter uniformity of the Si thickness is an advantage of SIMOX over bonded SOI, where the max to min variations are generally several times higher. This advantage becomes increasingly important as the average Si thickness is reduced.

For strained Si and SiGe, SE (and GXR) can be used to obtain layer thicknesses and compositions, and the optical constants determined by SE can be inputted into the Acumap software to create Si/SiGe material stack spectra libraries. This allows thickness maps of both the Si and SiGe films to be obtained. The downside is that new libraries must be constructed for each significant change in SiGe composition, and possibly for changes in relaxation as well. This will not represent a major problem once a composition and relaxation are "fixed" for development or manufacturing.

5.3   BURIED OXIDE INTEGRITY

The electrical properties of the buried oxide represented one of the earliest significant problems in SIMOX for both standard dose and later on for low-dose material. It was found that particles present on the surface just before the implant or particles generated during the implant would mask the Si surface to some degree, resulting in regions called pinholes which were oxygen deficient and caused high leakage or even short circuits through the BOX. The short circuit density of early material could easily range from 1 to 10 shorts per cm^2, destroying any yield probability for circuit chips that were generally $1\,cm^2$ or more in area. This problem has been largely overcome by extensive cleaning prior to loading the starting substrate into the implanter and removing the wafer after some portion of the implant has been completed, re-cleaning, and reloading before continuing the implant. Rotation of the wafer as part of this multiple re-cleaning also resulted in higher uniformity of the layer thicknesses by averaging out spacial non-uniformities in the implant.

The important electrical properties of the buried oxide that can affect the yield and/or reliability of SOI circuits are the short circuit density, the leakage current as a function of electric field, the breakdown voltage and field, the BOX charge, any trapping effects (particularly significant for radiation hardness), and any early breakdown events known as "mini-breakdowns".

Short circuit densities caused by particles have already been discussed. With modern techniques, the density of short circuits has been reduced to 0.1 per cm^2 or less, and most of these seem to occur near the wafer edges where they are less significant to circuit yield.

Leakage currents through the intrinsic BOX (i.e. leakages not caused by specific defects) follow well known electrical behaviour for insulating dielectric materials. Discussions of the I–V behaviour of buried oxides in SOI have been given by many authors in various issues of the Conference Proceedings of the IEEE International SOI Conference. At low fields, the conduction is ohmic due to the small density of mobile carriers. The resistivity of the BOX is 10^{12} to 10^{14} ohm cm at these low voltages. As the voltage is increased, injection of excess carriers over the Si/BOX energy barriers takes place accompanied by space charge limited conduction. At still higher fields, hopping of carriers or multi-step tunnelling through traps takes place, followed by Fowler–Nordheim tunnelling. When the dielectric strength is exceeded, irreversible oxide breakdown takes place. A schematic of the I–V behaviour in SIMOX buried oxides is shown in FIGURE 5.9.

For device and circuit purposes, it is the leakage currents associated with traps and other defect states at intermediate fields that are significant, since these can occur at low voltages where device operation takes place. Characterisation of leakage currents and breakdown behaviour is monitored using test structures with metal electrodes on the surface of the Si, followed by etching to isolate the individual devices. The simple test structure is shown in FIGURE 5.10. This structure is preferred for testing the I–V behaviour compared to deposition of the electrode directly on the BOX because removing the Si layer to expose the BOX has been found to etch away some of the relevant defects being studied. Positive voltages applied to the top electrodes with respect to the substrate cause depletion and eventual inversion in the p-type

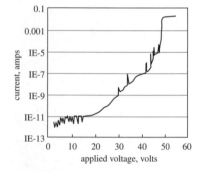

FIGURE 5.9 Log I–V characteristics of a buried oxide showing mini-breakdowns. Currents for voltages below 15 V are measurement noise.

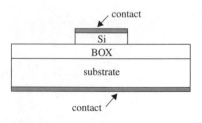

FIGURE 5.10 Device structure for buried oxide electrical measurements.

substrate and result in electrons injected into the oxide from the substrate. Negative voltages accumulate the substrate and result in electrons injected into the oxide from the Si film. Leakage currents are most conveniently measured at some fixed electric field for comparisons between wafers. Generally, 200 to 300 devices are measured across a quarter wafer test sample for statistical purposes. Typical leakage currents are 20 to 100 picoamps for fields of 2 megavolts per cm, which corresponds to voltages for low-dose material of 25 to 30 volts, well above any CMOS circuit operating regime. One possible long term effect of the traps, however, may be to accumulate charge and shift the threshold voltage.

The final, irreversible breakdown of the SIMOX buried oxides takes place at fields of 5 to 8 megavolts per cm, corresponding to voltages of 70 to 100 volts. The breakdown field increases as the implant dose is reduced at a fixed implant energy because the oxide becomes more stoichiometric with fewer amorphous Si inclusions and fewer Si islands (FIGURE 5.3). The use of ITOX (internal thermal oxidation) as part of the fabrication process also increases the breakdown fields. ITOX increases the thickness of the implanted BOX by diffusion of oxygen through the Si layer. This extra oxide at the top of the BOX is of higher quality with fewer defects, similar to thermal oxides, compared to the less stoichiometric buried oxide created by the implant.

Traps and their consequent effects at low voltages may pose questions for long term reliability, but the yield is much more determined by the mini-breakdown events which can cause fails during processing. These are breakdowns which take place at much lower voltages than the final, intrinsic breakdown. They show up as a sudden spike in the current, and are caused by field concentration due to the Si islands residing within the BOX (FIGURE 5.3). Since the dielectric constant of Si is substantially higher than that of SiO_2, the field is higher outside the Si islands than in the intrinsic oxide. The precipitates follow a normal (Gaussian-like) size distribution, with the largest islands equal to about half the BOX thickness, and the breakdown voltages near the islands can be half the intrinsic value. Once breakdown occurs, however, a current leakage path is created with very small dimensions and thermal runaway takes place, literally melting or vaporising the material in a small column around the precipitate so that the region becomes insulating again. These premature breakdown events are therefore self-healing.

The curve of FIGURE 5.9 shows how the I–V scan proceeds. At low voltages, currents of 0.1 to 10 picoamps are typical, although in this region the curve can be quite noisy. The current then follows the injection, space-charge limited flow, hopping conduction trend mentioned earlier until the first of these "mini-breakdowns" occurs for the largest islands. When this self-heals, the I–V curve continues until the next mini-breakdown is reached, and so on until all the Si islands in the test device have been vaporised. Each event is accompanied by a flash of light due to the high temperatures involved in the thermal runaway. Watching the test device under a microscope during the I–V scan reveals a "light show" with thousands of light flashes across the metal electrode until the last weak spot is gone.

Next to the short circuit density, the mini-breakdowns are the most important parameter to characterise for quality control of SOI buried oxides. If the mini-breakdown voltages are too low, damage to the material can take place during high voltage processes such as RIE, sputtering, plasma etching or plasma deposition. Circuit yields were seen to improve substantially as the implant dose and anneal conditions were tailored to reduce the Si precipitate size and density and increase the mini-breakdown voltages. FIGURE 5.11 shows a voltage scan of a low-dose SIMOX wafer on a linear scale. The current transients can easily exceed several orders of magnitude above the background and last for nanoseconds to microseconds. In order to prevent yield and reliability problems, both the first mini-breakdown event and the distribution of breakdowns should take place at voltages well above any expected during SIMOX processing. As long as the Si islands exist, consideration must be given to these breakdowns in any future technologies where the BOX and Si layer thicknesses are reduced. Another problem can arise if the BOX becomes thinner in any small area during processing, such as during STI formation. Once the BOX thickness becomes comparable to the largest Si island sizes, the breakdown voltages in these areas can be reduced to a few tens of volts and yield is severely impacted.

Mini-breakdowns are largely absent in bonded SOI wafers with their thermal buried oxides. However, occasionally such premature breakdowns take place even in bonded material, although at higher voltages than in SIMOX. The BOX in bonded material does not contain Si precipitates, but the cleaning steps, bonding procedure, bond anneals and possibly other processing steps may give rise to localised weak areas where the breakdown voltage is less than intrinsic. For both bonded material and SIMOX commercially available today, mini-breakdown voltages are usually sufficiently high to prevent yield losses during processing. However, the BOX leakage and mini-breakdowns should still be monitored for good quality control along with the pinhole density, since each new SOI material development such as Si/SiGe, SIMOX with lower implant energies and doses, 300 mm SIMOX, etc., can resurrect old problems or generate new ones.

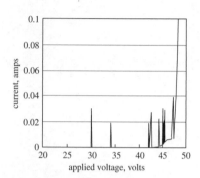

FIGURE 5.11   Linear I–V characteristics of a buried oxide showing mini-breakdowns.

5.4 SI LAYER ELECTRICAL PROPERTIES

Characterising the electrical properties of the Si layer is a major part of SOI substrate quality control. It is expected that the electrical properties of the virgin Si layer have some correlation with the electrical parameters after the substrate has undergone device and circuit fabrication (though this has never been proven). Techniques exist which can determine many key electrical parameters, even though the Si layer is nominally undoped and fully depleted. These parameters include the electron and hole mobilities, the subthreshold slope for both holes and electrons, the flat band and threshold voltages, the interface state density, the BOX charge, the saturation behaviour including drive currents, transconductances and output conductances, the mobility dependence on electric field, the doping level, and the generation and recombination lifetimes. Being able to determine these parameters in the virgin material saves a great deal of time and expense, since the alternative is to obtain them after circuit processing requiring a minimum of weeks to months. In contrast, characterisation of the starting material can be done in 1 to 2 days with virtually no processing at all.

One of the best tools for characterising the electrical properties of SOI starting material is the pseudo-FET [9]. This device takes advantage of the buried oxide acting as a gate oxide, allowing both inversion and accumulation layers to be created easily in both the film and the substrate. The device is made with two contacts to the Si layer, one acting as the source and one as the drain, with the substrate and BOX acting as the gate electrode and gate oxide. One version of a pseudo-FET is shown in FIGURE 5.12, in which the drain electrode in the centre is completely surrounded by a source electrode. The electrodes can be made with mercury or evaporated metal ohmic contacts [10] or with point contacts [9] similar to those used with 4-point probes. Pseudo-FETs made with mercury are termed HgFETs and those made with evaporated contacts are usually called RingFETs. They can also be made in a Hall-bar rectangular geometry where they are called HallFETs.

FIGURE 5.13 shows the subthreshold drain current–gate voltage (I_D–V_G) characteristics for both holes and electrons for HgFETs made on SIMOX with 120 nm Si films on a 140 nm BOX. Since the substrate is commonly p-type and the film is believed to be also (although it is possible the film is very slightly n-type), negative gate voltages with respect to the grounded source produce depleted and/or inverted substrate/BOX interfaces and accumulated Si films where hole currents dominate. Positive voltages on the gate produce accumulated substrates and inverted films where

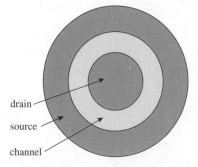

drain

source

channel

FIGURE 5.12   Geometry of the HgFET device for measurement of unprocessed starting material.

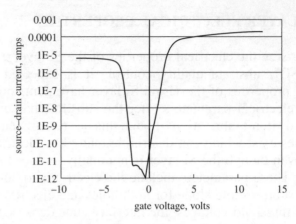

FIGURE 5.13   Drain current–gate voltage linear region I–V behaviour for an HgFET device on SIMOX. $V_D = 0.2$ V.

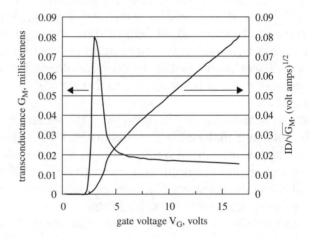

FIGURE 5.14   Transconductance and transfer curves from the data of FIGURE 5.13.

electron currents dominate. FIGURE 5.14 shows the transconductances (G_M) corresponding to these drain currents for electron flow in the film. The most prominent feature is the strong peak in G_M at voltages near the origin. The transconductance in a conventional FET does peak just above threshold, but in the HgFET the peak is exaggerated by the Hg electrode which is a leaky Schottky barrier rather than an ohmic contact.

The flatband and inversion threshold voltages are defined by convention as the intercept on the voltage axis of the tangent to the drain current taken around the peaks in G_M. The flatband voltage is the intercept at negative gate voltages, and the threshold voltage is the intercept at positive ones. To use these voltages in extracting the electrical properties of the Si film, they must be corrected for the voltage drop in the substrate [10], i.e. the gate electrode, since it is a Si "electrode" rather than a metal. The corrected value of the flatband voltage is used to obtain the fixed charge in the BOX:

$$QBOX = -\frac{(V_{FB} - W_{MS})\, C_{OX}}{Q} \tag{5.2}$$

where V_{FB} is the flatband voltage, C_{OX} is the buried oxide capacitance given by ε_{OX}/T_{OX}, and Q is the electron charge. W_{MS} is the difference in Fermi levels (work functions) between the Si film and the Si substrate, and since both are very low doped, $W_{MS} \approx 0$ in this case. Dividing by the electronic charge Q changes the units of QBOX from coulombs to a density, and both QBOX and C_{OX} have units of per cm^2.

The BOX charge is an important quality control variable. It affects the equilibrium surface potential in the Si film and therefore affects the threshold voltage as well as the flatband voltage. Low values of QBOX are desirable to minimise this effect on V_{TH}, and uniformity across the wafer is also desirable. Typical values for the BOX charge are 1.0 to 3.5×10^{11} per cm^2 (values are commonly reported as a density of states rather than a Coulombic charge) in low-dose SIMOX. At these densities, the effect of the BOX charge should be minimal and controllable. However, some processes for fabricating both SIMOX and bonded material with thin Si films have resulted in much higher densities, as much as 5 to 10×10^{11} per cm^2, which would be more of a concern for V_{TH} design and control. For SIMOX, the BOX charge may be related to the excess Si as shown in FIGURE 5.3. For bonded SOI, the charge may represent foreign material incorporated at the bonding seam, or a residual effect of hydrogen implantation.

The drain currents in FIGURE 5.13 can be analysed using the carrier mobility, the carrier density, and the geometry (width W and length L) of the device:

$$I_D = \frac{W}{L} q\, C_{OX} \frac{\mu_0}{1 + \theta(V_G - V_{TH} - V_D/2)}$$
$$\times \left(V_G - V_{TH} - \frac{V_D}{2} \right) V_D \tag{5.3}$$

where μ_0 is the low field mobility, V_D is the voltage from drain to the source, and θ is a surface scattering term which defines how rapidly the mobility decreases with increasing gate voltage. The transconductance in FIGURE 5.14 is the derivative of the current with respect to gate voltage. Using EQN (5.3), we can show that

TABLE 5.2   Electrical parameters for SIMOX wafers with different Si film thicknesses.

Si thickness, nm	Mobility (e)	Mobility (h)	QBOX $\times 10^{11}$	D_{IT} $\times 10^{11}$	SSL volts/decade	V_{TH} volt
117–160	606, 64	225, 29	1.5, 0.66	1.27, 0.46	0.27, 0.029	2.3–2.8
65–75	618, 60	238, 22	2.5, 0.91	2.55, 0.97	0.536, 0.08	4.5–5.5
53–57	635, 29	235, 13	2.7, 0.71	2.9, 0.71	0.573, 0.04	6.2–6.8

the drain current and the transconductance taken together can be used to obtain two important material parameters:

$$\frac{I_D}{\sqrt{G_M}} = \left(\frac{C_{OX} V_D \mu_0 W}{L} \right)^{1/2} \left(V_G - V_{TH} - \frac{V_D}{2} \right) \qquad (5.4)$$

An example of a curve of $I_D/\sqrt{G_M}$ versus gate voltage is also shown in FIGURE 5.14. The slope of this line yields the low field mobility and the intercept is a second measure of the threshold voltage. For pseudo-FETs with ohmic contacts, the two measures of V_{TH} agree very well. The same relationship applies for hole currents, and the hole mobility and accumulation threshold voltage (=flatband voltage) are obtained the same way.

The low field mobility averages and standard deviations for SIMOX wafers with various Si thicknesses are shown in TABLE 5.2. Each set of data represents 50 to 175 wafer lots. The average mobilities as shown are independent of the Si thickness. Other measurements on single SIMOX wafers where devices were made with different Si thicknesses by partial etching of the Si film confirm this independence. This is to be expected, since the inversion layer where the electron flow takes place is only 5 to 10 nm wide near the bottom interface. Average mobilities for electrons range from 610 to 640 cm^2 per V s. However, mobilities above 700 cm^2 per V s have been seen in a few cases. For holes, typical mobilities range from 210 to 240, but values as high as 270 have been observed. A histogram of electron mobilities in SIMOX wafers is shown in FIGURE 5.15 for low-dose material with 120 nm Si films. Most of the scatter in mobility values in FIGURE 5.15 comes from the wide range of conditions used to produce these SOI wafers during material development. Once a process is defined and reproduced in manufacturing, the histogram of mobilities should be much narrower.

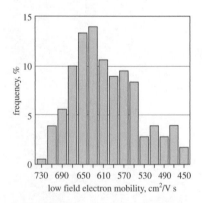

FIGURE 5.15   Low field electron mobilities for about 200 SIMOX low-dose wafer lots.

The subthreshold slopes can be used to determine the density of interface states at the Si/BOX interface and the doping level in the film. The subthreshold slope is given by:

$$SSL = \frac{kT \ln(10)}{QC_{OX}} (C_{OX} + C_{IT1} + C_{Si}) \tag{5.5}$$

where $C_{IT1} = QD_{IT1}$ is determined by the interface state density, $C_{Si} = \varepsilon_{Si}/T_{Si}$ is the "film capacitance" determined by the Si thickness, T is the temperature, and k is Boltzmann's constant. The subthreshold slope increases as the Si thickness is reduced. In conventional FETs, the gate oxide capacitance C_{OX} is much greater than C_{Si} or C_{IT1} and SSL approaches the ideal value of 60 millivolts per decade. For the thick buried oxides of SOI, typical subthreshold slopes are 0.25 to 0.5 volts per decade.

The inversion threshold voltage is given by:

$$V_{TH} = V_{FB} + \frac{2\phi_F}{C_{OX}} (C_{OX} + C_{IT1} + C_{Si}) + \frac{Q(N_A - N_D)T_{Si}}{2C_{OX}} \tag{5.6}$$

where $\phi_F = (kT/Q) \ln(N_A - N_D)/N_I$ is the "bulk potential" and N_I is the intrinsic carrier concentration. From EQNS (5.5) and (5.6), it is seen how the interface state density and the doping level are easily obtained from the measured flatband voltage, threshold voltage and subthreshold slope. Typical interface state densities range from 1.5 to 4×10^{11} per $cm^2 eV$, while typical doping levels for SIMOX are 2 to 5×10^{15} per cm^3. However, just as the BOX charge can show variability, the interface state density at the BOX/Si film interface can exhibit a wide range, with values from 5 to 10×10^{11} per cm^2 being observed in some cases, especially as the Si film has become thinner.

The interface state density and doping level can be found using gated diode and charge pumping measurements, in which MOS devices with $N^+/P/P^+$ junction regions are pulsed with various voltages on the drain and gate [4, 11]. Such measurements are more accurate than the pseudo-FET and show the variation of D_{IT} versus energy within the bandgap rather than the average value given by EQN (5.5). However, considerably more processing is required to fabricate the gated diode devices. The values of D_{IT1} given by these measurements are 10^{11}–10^{12} per $cm^2 eV$ [4, 11], in the same range as determined by the pseudo-FETs.

TABLE 5.2 shows typical values for these other Si layer parameters such as BOX charge, subthreshold slope, interface state density and threshold voltage, as well as hole and electron mobilities for low-dose SIMOX substrates.

The saturation I_D–V_D curves can be used to obtain other electrical parameters for the Si film. FIGURE 5.16 shows the saturation inversion layer I–V curves for electrons in a SIMOX wafer with a 120 nm Si film and 140 nm BOX. The slope of the current in the saturation region is the output conductance. In short channel FETs, the output conductance is a measure of the short channel effects such as DIBL (drain induced barrier lowering), the penetration of the drain depletion region into the channel. In long channel devices such as the pseudo-FET, the output conductance is more related to defects in the material or to a parasitic conducting channel. Saturation curves such as FIGURE 5.16 are obtained for both holes and electrons. Accumulation region I–V curves for holes are shown in FIGURE 5.17.

Another measure of the low field mobility can be obtained from the linear region of the saturation curves in FIGURES 5.15 and 5.16 near zero applied volts. The effective mobility is obtained at each value of gate voltage using the relationship:

$$\mu_{EFF} = \frac{L(\Delta I_D / \Delta V_D)}{WC_{OX}(V_G - V_{TH} - V_D/2)} \qquad (5.7)$$

where $\Delta I_D / \Delta V_D$ is the low voltage output conductance. FIGURE 5.18 shows a plot of μ_{EFF} versus $(V_G - V_{TH} - V_D/2)$. The intercept gives the low field mobility and is in good agreement with the values obtained from the I_D–V_G analysis using EQN (5.4). Another term available from the saturation I–V curves is the saturation mobility, assuming velocity saturation has not been reached which is a good assumption in long channel devices like pseudo-FETs. The saturation mobility is given by:

$$\mu_{SAT} = \frac{2I_{DSAT}}{(W/L)\,C_{OX}(V_G - V_{TSAT})^2} \qquad (5.8)$$

where I_{DSAT} represents the saturation value of the drain current for a given value of gate voltage V_G and V_{TSAT} is the threshold voltage in saturation. μ_{SAT} is always slightly less than the effective mobility μ_{EFF} because both the drain voltage and gate voltage are high for the μ_{SAT} measurement. Both hole and electron mobilities are obtained by these relationships or their equivalents.

It can be seen from the preceding paragraphs that a considerable number of quality control parameters are available by use of a pseudo-FET. The device is very simple, and can be made with pressure contacts [9], liquid metals such as Hg [10], or evaporated ohmic metal contacts [10]. Its advantages are simplicity,

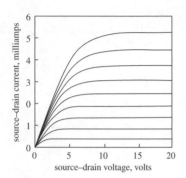

FIGURE 5.16 Drain current–drain voltage characteristics for a HgFET device on SIMOX for gate voltage steps of 1 V.

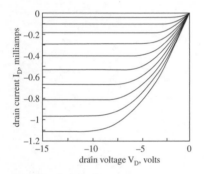

FIGURE 5.17 Drain current–drain voltage characteristics for holes measured by the HgFET device on SIMOX.

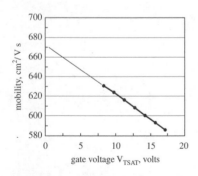

FIGURE 5.18 "Effective" mobilities versus gate voltage from the inversion layer saturation curves of FIGURE 5.16.

speed (fast turn-around), and low cost. The disadvantages are that it is not as accurate as some other techniques (full circuit processing), does not predict performance under the high doping levels present in finished devices, and cannot be used to study short channel effects. As a monitor for quality control of virgin SOI substrates and as a fast turnaround device for studying changes in material processing, its advantages far outweigh its disadvantages.

5.5   CAPACITANCE MEASUREMENTS

MOS capacitance measurements represent a powerful technique for studying bulk Si substrates and gate oxides. They can be used to a limited degree for characterising SOI substrates as well, particularly the buried oxide and the substrate below the BOX. FIGURE 5.19 shows a simple MOS capacitor made by depositing a metal electrode onto the buried oxide after removal of the Si film. (This is in contrast to the device of FIGURE 5.10 with the Si layer intact used to study BOX I–V behaviour.) FIGURE 5.20 shows a C–V measurement on a low-dose SIMOX wafer with an Au electrode. The two capacitance curves represent (1) capacitance measured at low frequencies where traps and interface states have sufficient time to follow the applied signal, and (2) capacitance measured at high frequencies where only the majority carriers can respond to the signal speed. The parameters available from the C–V measurement are the thickness of the oxide from the flat regions of the capacitance together with the device area, the BOX charge QBOX from the flatband voltage, the doping level from the flatband voltage and threshold voltage, the interface state density, the band bending versus applied voltage, and the leakage current at low voltages.

The flat band voltages and QBOX values obtained from MOS capacitors made on SIMOX after removal of the Si film are in excellent agreement with the values obtained from pseudo-FETs made on the material before removing the Si. The difference between the V_{FB} measured by the C–V method and that obtained from the pseudo-FET is that there is no correction in the C–V analysis for the surface potential drop in the substrate; the substrate surface potential is an integral part of the C–V measurement. The BOX charge is obtained from EQN (5.2) after dividing by the area to obtain a density per cm^2:

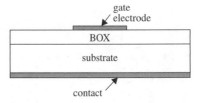

FIGURE 5.19   SIMOX wafer with the Si layer removed for BOX capacitor measurements.

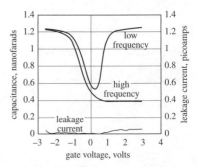

FIGURE 5.20   C–V and accompanying I–V (absolute magnitude) curves of typical BOX capacitance on SIMOX.

$$QBOX = \frac{C_{OX}(W_{MS} - V_{FB})}{Q \ Area}$$ (5.9)

95

where W_{MS} here is the work function difference between the substrate and the metal gate electrode.

The leakage current obtained from the C–V measurement is another indication of the quality of the BOX. The leakage current is obtained from the quasi-static capacitance measurement where the capacitance is obtained by integrating the displacement current that flows as the result of a voltage step. Once the displacement current has disappeared, the remaining current is then an "equilibrium" leakage current. As long as the device structure is as shown in FIGURE 5.19, where the BOX area is much larger than the electrode area so that there is no edge leakage from exposed device edges, and assuming low loss cables and controlled humidity, this leakage current flows through the BOX and is an indication of the traps and other possible defects in the BOX. High mobile charge densities could also contribute to the leakage, but high densities are unlikely in SIMOX and bonded SOI oxides. Figure 5.20 shows that leakage currents corresponding to well behaved capacitance measurements are in the range of fractions of a picoamp. If there is significant leakage current from any of these sources, the low frequency capacitance is strongly affected and cannot be used for parameter extraction. FIGURE 5.21 shows an MOS capacitance measurement on a different SIMOX wafer with higher leakage current, but even though the current is only 1 picoamp or less, it causes curvature in the C–V curves which makes them unusable.

Since so much attention has to be paid to minimising leakage currents at low voltages in the C–V measurement, much smaller currents can be studied by the C–V technique than in the BOX I–V measurements discussed earlier. For example, the currents of the two devices of FIGURES 5.20 and 5.21 are in the femtoamp to picoamp range, while the currents in the BOX I–V measurement shown in FIGURE 5.9 are mostly noise below 10 picoamps due to using commonly available parameter analysers. Of course, special low noise analysers and special shielding could be set up to measure very low leakage currents in the BOX I–V directly, but the low frequency ("quasi-static") C–V measurements give these leakage currents as a "bonus". These leakage currents occur in a voltage range (-3 to $+3$ volts is typical) at which the circuit devices made on these substrates will actually operate. As mentioned earlier, the leakage currents are related to defects in the BOX and are therefore a useful measure of the BOX quality. The quasi-static measurement is slow, however, caused by the time necessary for the displacement current to disappear. Twenty to thirty minutes is typical for the C–V measurement, compared to one to two minutes for the BOX I–V, and the BOX I–V measurement is carried out over a much larger voltage range.

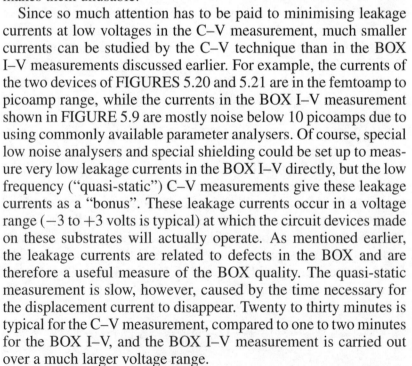

FIGURE 5.21 C–V and I–V curves for a BOX capacitor with substantial leakage current.

Another C–V measurement that can be made to study BOX quality is the C–V shift caused by a change in the charge profile in the BOX due to some stimulus. The C–V shift has been used in radiation hardness studies where fixed doses of radiation are incident on the substrate. The flatband voltage shift is a measure of the change in charge trapping caused by the radiation: the less flatband shift, the more radiation tolerant the device is. Shifts in the C–V profile can also be used to study defect densities in the BOX by flowing a fixed current for some length of time ("current stress") and examining the change in flatband voltage. Measurements carried out on standard dose SIMOX (400 nm BOX thicknesses) showed that a 100 nanoamp per cm^2 current density for 10 minutes could shift the flatband voltage by -15 volts. Both bonded SOI and low-dose SIMOX have much smaller shifts. Bonded SOI oxides are thermal oxides and exhibit fewer traps. Low-dose SIMOX made with the ITOX anneal process also has more thermal-like oxides and fewer traps.

For low-dose material, a second effect can take place as a result of current stress measurements: a greatly increased BOX leakage current at low voltages, even when there is little shift in the flatband voltage. The leakage current can easily increase by several orders of magnitude after applying a fixed voltage of 20 to 30 volts for ten minutes, which greatly distorts the C–V curves and makes them unusable. C–V shift measurements using current flow stimulus on low-dose material are problematic and not as useful as studying the equilibrium current obtained during quasi-static C–V measurements. Studies of C–V shifts from radiation exposure, on the other hand, would be very useful.

Other parameters such as D_{IT} and doping obtained by C–V measurement on BOX capacitors such as FIGURE 5.19 are of only marginal use in SIMOX quality control. The reason is that the doping level and interface state density are those of the substrate rather than the Si film where the circuits will be made. Of course, MOS capacitors can be made to SOI material with the Si film left on, as shown in the device in FIGURE 5.10, known as an S-I-S structure. However, there are now at least two capacitors in series: the Si film/BOX and the BOX/substrate. A third or fourth capacitor may be present if the metal electrodes to the Si film and substrate are not ohmic. The device acts as a voltage divider for the applied voltage, with an unknown value of voltage falling across the different regions. Interpretation of the measurements to extract parameters applying to the Si film are considerably more difficult than interpretation of the BOX capacitor of FIGURE 5.19. Cristoloveanu and Li [4] have described the physics and parameter extraction of S-I-S capacitors, assuming ohmic contacts at the two Si surfaces. Parameters such as interface state densities can be extracted using coupling equations between the two BOX interfaces. C–V curves appear to be inverted U-shaped, with a peak near zero volts and decreased capacitance for either higher or lower voltages.

C–V measurements of standard dose SIMOX S-I-S capacitors with 200 nm Si films follow this expected U-shaped trend. However, measurements of low-dose p-type S-I-S capacitors with thin Si films generally show modified behaviour. The C–V behaviour of a normal BOX capacitor has already been shown in FIGURE 5.20. FIGURE 5.22 shows an S-I-S capacitor on the same substrate, using Au electrodes as ohmic contacts on both surfaces. The S-I-S behaviour has a generally similar appearance to that of the BOX capacitor, except for a slight shift in V_{FB} and a dramatic shift in the high frequency capacitance for positive voltage polarity. In other cases, the C–V curves of S-I-S devices on low-dose material are nearly identical to those of BOX capacitors made on the same substrate. FIGURE 5.23 shows another S-I-S capacitor on the same substrate as FIGURE 5.22, demonstrating that C–V curves very similar to those of BOX capacitors can be obtained with S-I-S devices under some conditions. This suggests that one of the series capacitors in the S-I-S device becomes inactive under certain conditions, perhaps by being shunted by a high leakage current. More work is needed to understand what these effects are in thin BOX, thin Si material.

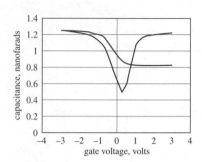

FIGURE 5.22   C–V plot of an Au/Si/BOX capacitor on SIMOX, showing the effect of multiple series capacitors.

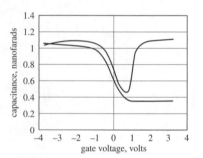

FIGURE 5.23   C–V plot of another Au/BOX capacitor on the same SIMOX wafer as in FIGURE 5.22.

5.6   GATE OXIDE INTEGRITY

Gate oxide integrity is a measurement of the electrical quality of the gate oxide used to form the MOSFET. In order to produce the highest performance in the device, the gate oxide must sustain fields close to its dielectric breakdown strength, should have an acceptably low density of fixed charge, and should be able to withstand many millions of switching cycles. GOI has been studied at great length in bulk Si MOSFETs. Techniques include leakage current, breakdown field, density of shorts and charge-to-breakdown. Accelerated "stress" tests are carried out at elevated temperatures with high applied fields. Statistics on these parameters are used to judge the oxide quality. C–V measurements are useful in studying leakage currents and oxide charge. I–V measurements are useful for leakage currents, breakdown voltages and charge-to-breakdown.

By comparison with bulk material, much less study of GOI on SOI material has been done. This may be due in part to the fact that the same causes of poor GOI in bulk material should be applicable in SOI: high metal densities, high particle count and defects of various kinds. One factor in addition present for SOI material is surface roughness. For bulk Si, the surface roughness is generally 1 angstrom or less RMS, while for SIMOX the roughness is 7 to 9 angstroms for standard dose and 3 to 5 angstroms for low-dose. Bonded material, which generally undergoes a smoothing polish step, often exhibits roughness of 1 to 2 angstroms. A conformal gate oxide several times thicker than the surface roughness undulations overlaying a Si surface is probably not affected by the roughness, and even modern gate oxides of 10 to 15 angstroms have not exhibited problems with GOI for present circuit generations.

Measurements of GOI for bulk Si, standard dose SIMOX and low-dose SIMOX with 5 nm oxides have shown that the GOI is equivalent in the three materials. FIGURE 5.24 shows the breakdown voltages for the first 100 out of 700 devices made by evaporating $0.05\,cm^2$ gold mesas on the 5 nm oxides on standard dose SIMOX. The SOI devices were made by ion implanting and annealing an n^+ region into the Si film followed by ohmic contact formation and the Au mesa evaporation. The breakdown voltages of 3.5 to 3.8 volts are close to the breakdown field strength of 8 to 10 megavolts per cm for good quality oxides and are identical to the breakdown voltages of control bulk Si monitors. The breakdown voltages of 100 devices made on 4×10^{17} per cm^2 implant dose SIMOX are shown in FIGURE 5.25. Early low-dose material had many more surface pits, mounds, metal contamination and possibly other defects which showed up as variability in the gate oxide integrity: low breakdown voltages (as seen in FIGURE 5.25) and occasionally short circuits. Modern vintage low-dose material

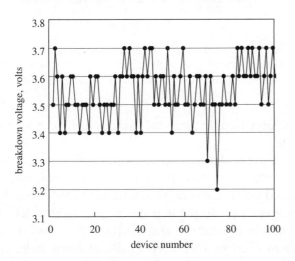

FIGURE 5.24   Breakdown voltages for 100 Au/gate oxide/Si capacitors on standard dose SIMOX. Gate oxide thickness = 5 nm.

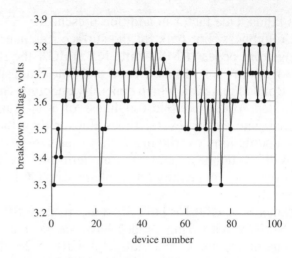

FIGURE 5.25　Breakdown voltages for 100 Au/gate oxide/Si capacitors on low-dose SIMOX. Gate oxide thickness $= 5\,\text{nm}$.

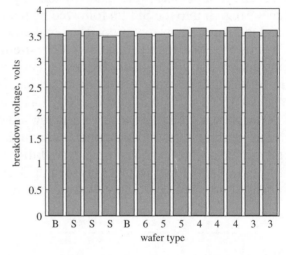

FIGURE 5.26　Mean breakdown voltage comparison for 700 devices on each wafer. S $=$ standard dose; $6 = 6 \times 10^{17}$; $5 = 5 \times 10^{17}$; $4 = 4 \times 10^{17}$; $3 = 3 \times 10^{17}\,\text{cm}^{-2}$ implant dose. Gate oxide thickness $= 5\,\text{nm}$.

has greatly improved surfaces and Si film quality, with consequent higher GOI.

The average breakdown voltages for 700 Au mesa devices on 5 nm gate oxides are shown in FIGURE 5.26, where bulk Si wafers are compared against several standard dose SIMOX and low-dose SIMOX wafers with implanted doses of 3 to 6×10^{17} per cm^2. It is clear that bulk material and SIMOX of any implant dose are equivalent in GOI as measured by breakdown field, at least at these oxide thicknesses. Leakage currents for 25 devices made on each of these wafers were also statistically equivalent. GOI measurements on bonded SOI also showed it to be equivalent to bulk silicon and SIMOX. These data support the suggestion that GOI should be affected by the same factors whether on bulk material or SOI, no better and no worse. The same should be true for the high-K dielectrics to be used as gate insulators for the next generations of CMOS. Films of O-N-O (oxide-nitride-oxide) and higher dielectric constant films such as HfO_2 and the like are expected to behave the same on SOI as on bulk.

5.7 LIFETIME

The carrier lifetimes in semiconductor wafers are valuable measures of the material quality, since point defects, surfaces, dislocations, stacking faults and other material defects all impact on the lifetime value. The presence of metallic impurities such as Fe, Cu, Ni, Au and others can have a strong effect on the lifetime, and the recombination lifetime (or diffusion length) can often be taken as a measure of the metal impurity density.

The same two lifetimes applicable to bulk Si are also important in SOI material. Recombination lifetime is a measure of the rate at which excess holes and electrons recombine with each other so that the sample returns to equilibrium. Generation lifetime is a measure of the rate at which electron–hole pairs are created in response to a carrier deficiency, again in order to return to equilibrium. The counterparts to these two lifetimes at surfaces and interfaces are known as the surface recombination velocity (SRV) and surface generation velocity (SGV), which are of very high importance in SOI material. The "effective" lifetimes are a combination of bulk and surface terms:

$$\frac{1}{\tau_{EFF}} = \frac{1}{\tau_{BULK}} + \frac{1}{\tau_{SURFACE}} \quad (5.10)$$

In bulk Si, τ_{EFF} can approach τ_{BULK}, although the surface can still have a significant effect. In SOI, the surface term dominates.

Techniques for measuring lifetimes in Si (and in some cases other materials) include surface photovoltage, free carrier absorption, photoconductivity decay, electron beam induced current, pulsed MOS capacitors, gated diode, diode recovery, photoluminescence decay and corona discharge, among others. Most of these can be used to examine the bulk Si substrate beneath the BOX layer in SOI material, but are problematic for measuring the lifetimes in the thin Si film. An excellent discussion of these various techniques can be found in [5].

5.7.1 Photovoltage

Surface photovoltage is a widely used technique for measuring bulk Si lifetimes and for relating these lifetimes to heavy metal impurity contamination. In this method, carriers are excited at several different wavelengths, often with the use of narrow bandpass optical filters. The change in surface potential due to the modulated incident light is detected capacitively by a pick-up probe placed near the surface. A plot of incident photon flux versus absorption depth (determined by the wavelength) for constant photovoltage, or alternatively the photovoltage versus absorption depth for constant photon flux, determines the minority carrier diffusion length, related to the lifetime by $\tau = L^2/D$ where D is the diffusion constant. Photovoltage measurements have been carried out in SIMOX by shining light on the backside of the wafer, and on the wafer front after removal of the Si film and BOX [12]. The diffusion lengths were essentially the same, which is attributed to the high diffusion coefficients of heavy metal impurities which homogenise throughout the bulk during the anneals. The diffusion lengths have been correlated with impurity densities, particularly iron [5, 12] (provided this is the dominant recombination centre). Several commercial tools are available for performing these measurements. In one instrument, known as ELYMAT, a bulk wafer is immersed in 1–2% HF solution, which passivates the surface, minimises the surface recombination portion, and gives the closest answer to the true bulk lifetime, as long as the sample is several times thicker than the minority carrier diffusion length.

Unfortunately, surface photovoltage cannot determine the lifetime in the Si film. The technique inherently requires the thickness of the measured region to be at least as thick as a diffusion length and preferably several diffusion lengths. The typical Si film thickness of 100–200 nm is far less than the typical Si diffusion lengths of 300–600 microns. In addition, surface recombination dominates in such thin Si films, even for passivated surfaces. Moreover, the Si film must be removed even to measure the photovoltage in the bulk substrate beneath the BOX, since the Si/BOX, BOX, and BOX/substrate interface all represent capacitances which interfere with the capacitive pickup of the photovoltage signal. The ELYMAT tool, however, could be used to examine SOI wafers by shining light and making measurements on the backside of the wafer.

5.7.2 Excess carrier decay

The creation and detection of excess carriers generated by a pulse of light is a powerful technique for measuring the recombination lifetime. The carriers are created by a "pump" beam which can be light of a single wavelength such as a laser or a narrow range of wavelengths using a narrow bandpass filter, or it can be broadband "white" light, for example from a xenon flashlamp. The essential feature is that the photon energy exceeds the semiconductor bandgap. Detection can be carried out in several ways. If ohmic contacts are placed across the sample, the photocurrent can be obtained directly. Lock-in amplifier techniques can be used to detect photoconductances which are much smaller than the background, allowing measurements to be made at low injection levels. An alternative technique which is non-contact is to use a "probe" beam transmitted through the sample to detect the change in free carrier absorption due to the generated photocarriers [13]. The probe beam may be an infrared laser or a focused black body source. Fast infrared detectors are used to monitor the temporal decay of the generated carriers after the end of the light pulse with several nanosecond resolution. Modulation techniques allow measurements at low injection levels as in the contacted device.

A third technique which has become widely used is photoconductivity (PC) decay using microwaves reflected from the wafer for detection. The microwave reflectivity is a strong function of the sample conductance in a tuned cavity; therefore the microwave reflectance directly mirrors the density of photogenerated carriers if the mobility is constant, and the lifetimes determined by infrared transmittance and microwave reflectance are the same at equal injection levels. However, the PC decay method generally involves higher injection levels than the other techniques in order to have sufficient signal-to-noise discrimination over the substrate background.

The average density of carriers in a semiconductor slab of thickness T is given by [14]:

$$N_{av}(t) = \sum B_k b_k \frac{\sin z_k}{z_k} \exp\left[-\left(\frac{1}{\tau_b} + a_k^2 D\right)t\right] \quad (5.11)$$

where B_k, b_k and z_k are all terms containing the surface recombination velocities, sample thickness, or both, but not the time. D is the diffusion coefficient of the generated carriers. The coefficients a_k are given by [14]:

$$a_k T = \tan^{-1}\left(\frac{S_1}{Da_k}\right) + \tan^{-1}\left(\frac{S_2}{Da_k}\right) + k\pi \quad (5.12)$$

where the terms in EQN (5.12) and the summation in EQN (5.11) are over all integer values of k. S_1 and S_2 are the surface recombination velocities at the front and back surfaces, respectively. A commonly made assumption, though not always correct, is that the SRVs at the front and back surfaces are the same ($S_1 = S_2 = S$), which results in a simplified calculation of the time coefficients:

$$a_k \tan \left(\frac{a_k T}{2} \right) = \frac{S}{D} \qquad (5.13)$$

The time dependence of the decay due to the excess carriers is given by the exponential terms in EQN (5.11) with amplitudes given by the pre-exponential terms. To a large extent, the first value of a_k obtained for $k = 0$ is the most important, and the other terms for higher values of k die out quickly in the decay signal. The surface recombination time dependence is contained in a_k, which increases as the surface recombination values increase.

For SOI, the pulse of photogenerated carriers can be confined to the thin Si film by the use of an ultraviolet light pulse. The absorption coefficient of UV light in Si reaches and exceeds 10^6 per cm[1], so that the light is absorbed in several tens of nm of material. The carriers homogenise quickly throughout the film and are confined to the film by the high Si/BOX energy barrier. In almost all cases, the generated carriers recombine at the two surfaces rather than in the "bulk" of the film, as a calculation of the surface lifetime term in EQN (5.10) will show. If the surface recombination is low at both surfaces, 0.25 to 1 cm per s or less for carefully passivated surfaces [15], then the surface lifetime term in EQN (5.10) is given by $\tau_{SURFACE} = T/2S \approx 10$–40 microseconds for a 200 nm Si film. If the surfaces are not passivated, then the surface term is given approximately by $\tau_{SURFACE} = T^2/\pi^2 D$ which is 1–2 picoseconds for a 200 nm film. Since the bulk lifetime in Si is usually 100–300 microseconds in good quality material, the surface term dominates in either case. The film would have to contain a very high density of defects to lower the bulk lifetime below that of the surface. In general, a qualified gate oxide on the upper surface reduces the SRV at the top to a very small value so that the lifetime measurement in the Si film is really a measurement of recombination at the Si/BOX interface.

FIGURE 5.27 shows PC decay signals from several SIMOX wafers with Si films about 200 nm thick. The signals were generated with pulses from a nitrogen flashlamp with a 5 nanosecond pulse width and fall time. The measured effective lifetimes are typically from 200 to 500 nanoseconds in material with gate oxide surface passivation. Assuming these times represent back

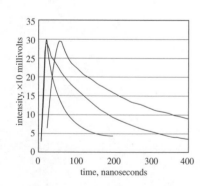

FIGURE 5.27 Photoconductivity decay in response to laser light pulse incident on Si layer surface. Lifetimes for the curves shown are 43, 118 and 312 nanoseconds.

surface recombination, the SRV value at the back surface is about 20–50 cm per second.

The substrate lifetime below the BOX can easily be measured either by removing the Si film by selective etching, or non-destructively by using longer wavelength light which penetrates through the Si film. The absorption coefficient decreases rapidly for wavelengths above 400 nm, so that the film absorbs very little of the incident light and virtually all the PC decay signal emanates from the bulk substrate. FIGURE 5.28 shows PC decay signals from the substrates in several SIMOX wafers with both high and low metal impurity levels as measured by SIMS. The high metal contamination lowers the lifetime by a factor of 2. Other metals such as Cu have an even larger effect on the lifetime, as shown in FIGURE 5.29. PC decay has been used to monitor the presence of iron [16] and other metal contamination in SOI material. It is clear that pulse-generated carrier decay measurements can be used like surface photovoltage to detect metal contamination in Si substrates.

The shape of the decay can determine whether the metal is located prevalently at the BOX interface or distributed evenly throughout the bulk. Recombination centres predominately at the interface cause a sharp, fast decrease of the decay signal immediately after the excitation pulse ends, while recombination centres distributed evenly throughout the material cause a more gradual decay. FIGURE 5.28 is an example of SOI material with metal distributed evenly, while FIGURE 5.29 is an example of metal located close to the BOX/substrate interface as measured by SIMS [17].

The ability to measure the effective lifetime for the film and the substrate below the BOX simply by changing the excitation wavelength is a valuable aspect of the PC decay technique. FIGURES 5.30 and 5.31 show the Si film and bulk lifetimes measured by microwave-PC decay for a half SIMOX wafer where the film lifetime was measured with the nitrogen flashlamp (wavelength 340 nm) and the bulk lifetime was measured with a xenon flashlamp with the Si film still on, demonstrating that PC decay can be a non-destructive tool for quality control of starting material. IR absorption decay could be used in a similar manner and would result in curves equivalent to FIGURES 5.28 and 5.29. FIGURE 5.31 shows that the substrate lifetimes in SIMOX are $\frac{1}{4}$ or less than typical values for virgin bulk material, indicative of residual metals, damage or other recombination centres as a result of the implant and anneal.

IR absorption by photogenerated carriers can also be used in another lifetime measurement technique by low frequency modulation of both a pump and a probe beam and signal detection

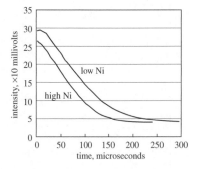

FIGURE 5.28 Photoconductivity decay in the substrate below the BOX for SIMOX wafers with low and high Ni content.

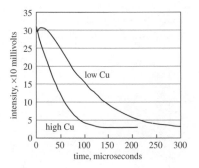

FIGURE 5.29 Photoconductivity decay for wafers with low or high Cu contamination near the BOX/substrate interface.

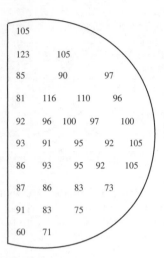

FIGURE 5.30 Photoconductivity decay lifetime in nanoseconds of a Si film of a SIMOX substrate.

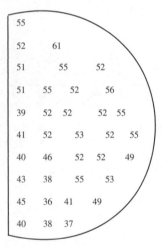

FIGURE 5.31 Photoconductivity decay lifetime in microseconds for the substrate below the BOX in the SIMOX wafer of FIGURE 5.30.

with a lok-in amplifier [18, 19]. Using alternately short and longer wavelength pump beams, estimates of the effective lifetimes in both the film and substrate regions can be obtained. Lifetimes of around 200 nanoseconds and 15–20 microseconds were reported for the film and substrate, respectively [18]. As discussed above, this film lifetime is nearly always a measure of the surface recombination velocity for films below 1 micron in thickness [19, 20].

The effective lifetimes in the Si films of both SIMOX and bonded wafers have also been measured by photoluminescence decay at 4.2 K for UV pulse excitation [21]. The photoluminescence efficiency is greatly increased by cooling the material to these temperatures. The measured film lifetimes of 80 to 165 nanoseconds for SIMOX and 320 nanoseconds for bonded wafers are similar to values obtained by the other methods discussed earlier.

5.7.3 Generation lifetime

The generation lifetime in SOI material is at least as interesting as the recombination lifetime and probably more so for several reasons. First, it is relatively straightforward to measure this lifetime in the thin Si film. Second, it is less affected by surface processes which dominate recombination. Third, the generation lifetime measurement has a natural magnification factor which causes the transients to have seconds to minutes temporal decay rather than nanoseconds. Fourth, it can be measured in either virgin starting material or finished devices.

The low injection level recombination lifetime in p-type material is approximated by [5]:

$$
\tau_{REC} = \tau_P \left(\frac{n_I}{N_A} \right) \exp \left[\frac{(E_R - E_I)}{kT} \right]
$$
$$
+ \frac{\tau_N}{N_A} \left(N_A + n_I \exp \left[-\frac{(E_R - E_I)}{kT} \right] \right)
$$
$$
\approx \tau_N \tag{5.14}
$$

where E_R is the energy level of the recombination centre. The lifetime τ_N is given by $(1/\sigma_N \, v_{TH} \, N_R)$, where σ_N is the capture cross section, v_{TH} is the thermal velocity and N_R is the density of recombination centres. The value of τ_{REC} is not greatly affected by the trap position E_R. The generation lifetime, on the other hand, is given by [5]:

$$
\tau_{GEN} = \tau_P \exp \left[\frac{(E_R - E_I)}{kT} \right] + \tau_N \exp \left[-\frac{(E_R - E_I)}{kT} \right] \tag{5.15}
$$

and is greatly affected by the trap position E_R. For trap positions several kT removed from the intrinsic Fermi level E_I, the generation lifetime can be 10 to 100 times the value of τ_{REC} [5].

Similarly, the surface recombination velocity for low level injection is given by:

$$S_{REC} = \frac{S_N S_P N_A}{S_N \, n_I \exp[(E_{INT} - E_I)/kT] + S_P[N_A + n_I \exp[-(E_{INT} - E_I)/kT]]}$$

$$\approx S_N \quad \text{(p-type material)} \tag{5.16}$$

On the other hand, the surface generation velocity in low level injection is given by:

$$S_{GEN} = \frac{S_N S_P}{S_N \exp[(E_{INT} - E_I)/kT] + S_P \exp[-(E_{INT} - E_I)/kT]} \tag{5.17}$$

$$\approx S_P \exp\left[-\frac{(E_{INT} - E_I)}{kT} \right]$$

for $E_{INT} > E_I$ by several kT, or

$$\approx S_N \exp\left[-\frac{(E_I - E_{INT})}{kT} \right]$$

for $E_I > E_{INT}$ by several kT

where E_{INT} is the trap energy position at the interface. EQNS (5.16) and (5.17) show that the surface generation velocity can be considerably less than the recombination velocity for trap energies away from mid-gap.

The effective generation lifetime can be obtained on virgin SOI substrates without any processing at all using the pseudo-MOS device with point contacts [9, 22] or mercury contacts [10], or with simple evaporated metal electrodes. The test sample only requires two surface electrodes acting as source and drain and a bottom contact (on the substrate) acting as the gate electrode, with the BOX acting as the gate oxide; in other words, the same structure as the HgFET. With a small drain voltage applied, the gate electrode is abruptly switched on from the off state and the increasing drain current transient is monitored, or switched off from an on state and the drain current decay is monitored. FIGURE 5.32 shows current transients taken at room temperature for a SIMOX wafer using evaporated metal electrodes, Au for the hole current transient and

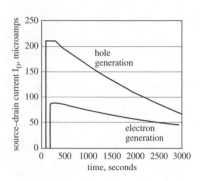

FIGURE 5.32 Electron and hole
current decay after removal of gate
voltages driving the device into
either accumulation or depletion.

ErAu for the electron current. Switching the gate voltage on produces the abrupt current increase, followed by the plateau region as long as the gate voltage is applied. Sudden removal of the gate voltage produces the slow decrease as shown. In FIGURE 5.32, the curve marked "hole generation" is taken with positive gate voltage which produces an electron current. The decay is then due to thermal generation of holes which compensate the electrons. The curve marked "electron generation" is a hole current produced by negative gate voltage, where the decay is caused by thermal generation of electrons. Using Hg electrodes is particularly convenient in that both hole and electron lifetimes can be obtained with the same device. The striking feature of these measurements is the time scale; the drain current persists from hundreds of seconds to several hours in high quality material. The decay time is a magnified representation of the generation lifetime, given by [5, 23]:

$$\tau_{\text{RECOVERY}} = (1 - 10)\left(\frac{N_A}{n_I}\right)\tau_{\text{GEN,EFF}} \tag{5.18}$$

where $\tau_{\text{GEN, EFF}}$ is the effective generation lifetime which includes any contribution from surface generation. Since N_A/n_I is more than 10^5 at room temperature for most SOI material, the transients such as those of FIGURE 5.32 are magnifications of about 10^6 of the generation lifetimes in SOI wafers. The estimated lifetimes for these wafers are in the millisecond range. A Zerbst-type analysis [5, 23] is needed to determine $\tau_{\text{GEN,EFF}}$ accurately. Previous calculations of early standard dose SIMOX indicated generation lifetimes of 130 microseconds [23], but SOI material has been greatly improved since then.

The decay of drain current in response to changes in gate voltage has more commonly been used with finished devices, including standard nMOS devices [23] and n^+-n $-n^+$ enhancement mode transistors [24]. Pulsed decay plots of

$$A\frac{d}{dt}[K(\Delta I_{DS} + B)]^2 \text{ vs } \left(\frac{K\,\Delta I_{DS}}{\tau_{\text{GEN}}}\right) + S_{\text{GEN}}$$

are the current analogue of the Zerbst plots using capacitance [5] where A, K and B are constants. The slope of the curve yields the generation lifetime and the intercept gives the surface generation velocity. Since the doping levels in finished devices are much higher than in virgin material, the generation lifetimes are lower. Lifetimes of 40 to 130 microseconds in the Si film have been reported in early SOI MOS transistors [23, 24] made with standard dose

SIMOX . Longer lifetimes can be expected in recent material such as low-dose SOI with improved defect densities.

The commonly used method for obtaining generation lifetimes and surface generation velocities in bulk Si is the time dependence of pulsed MOS capacitors. A gate voltage pulse is applied which biases the device from accumulation or weak depletion into strong depletion. The depletion width is larger than the normal depletion width in equilibrium, and the capacitance is smaller than the equilibrium capacitance. The inversion layer forms, and the capacitance relaxes back to equilibrium, by thermal generation of hole–electron pairs. The rate at which the relaxation occurs is a measure of the generation lifetime. Analysis of the capacitance transient, known as a Zerbst analysis, yields the generation lifetime and surface generation velocity.

It is clear that the pulsed capacitance method cannot be used in virgin SOI material, where the doping is low and the Si film is already fully depleted in equilibrium. If the device is doped high enough to be only partially depleted, the pulsed capacitance method can be used but is more difficult than the pulsed drain current method described above due to series resistance parasitics which complicate the capacitance measurement on thin Si films. This might explain why pulsed current transients have been used successfully on SOI material [23–25] but pulsed capacitance measurements of the Si film are seldom reported. Of course, pulsed capacitance measurements could be made on the SIMOX substrate region after removal of the Si film, yielding the generation lifetime of the bulk region.

Gated diode measurements have been used to examine generation lifetimes in SOI Si films, while charge pumping measurements on gated diodes have been used to determine interface state densities [11, 26, 27]. In gated diodes, N^+-N-P^+ or N^+-P-P^+ diodes are created laterally in the Si film and a gate oxide and gate contact overlap the two heavily doped regions. Gate voltages are applied ranging from less than flatband to higher than inversion threshold, resulting in both body and surface generation of electron–hole pairs, leading to a small current when the gate voltage lies between flatband and threshold, as shown in FIGURE 5.33. DC measurements using gated diodes have indicated generation lifetimes of tens of microseconds and surface generation velocities as low as 0.3–1.5 cm per s. In the charge pumping measurement, which can also be carried out with MOSFETs, an AC signal is applied to the gate with enough amplitude to drive the device alternately into accumulation and inversion. Holes and electrons are captured and emitted by interface states residing on one or both interfaces (Si/gate oxide and Si/BOX) during alternate phases of the gate voltage. The result is a current

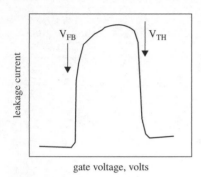

FIGURE 5.33   Typical I–V characteristics of a gated diode structure.

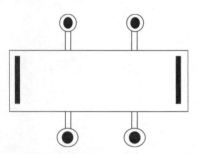

FIGURE 5.34   Geometry of the Hall bar used for Hall effect and HallFET measurements.

(the "charge pumping" current) which is very similar in features to the gated diode current of FIGURE 5.33 [26]. Interface state densities of $1–2 \times 10^{11}$ per cm^2eV^{-1} have been measured on SIMOX substrates, similar to the values obtained by pseudo-FET measurements on unprocessed material as described in Section 5.4.

5.8   OTHER ELECTRICAL TECHNIQUES

5.8.1   Hall effect

The Hall effect is an independent way of measuring hole and electron mobilities, sheet resistivities and sheet carrier densities as a function of doping, electric field, temperature or other relevant parameters. The Hall effect is usually measured using a rectangular sample, or "Hall bar", as shown in FIGURE 5.34, often called a Hall "spider" because of its shape. A small voltage is applied to the two end contacts so that a current is established and an ohmic voltage drop appears across the legs of the spider. Since no current flows into the legs, the voltage across the legs is an accurate measure of the voltage along the bar. The sheet resistivity is then given by:

$$\frac{\rho}{t} = \frac{V_S W}{L I_D} \tag{5.19}$$

where V_S is the voltage across the legs, t is the sample thickness, L is the distance between the legs, W is the width of the bar, and I_D is the current. The voltage is also measured across two opposite legs, and if the sample is geometrically perfect, this voltage would be negligible, but samples are rarely that perfect and a small voltage V_P will be measured. A magnetic field is then applied perpendicular to the surface, which creates a force that shifts the carriers toward one side of the bar. The change in voltage ΔV_P is known as the Hall voltage, and it is used to obtain the Hall mobility and the sheet carrier concentration using:

$$\mu_H = \frac{V_H}{I_D B \rho / t} \qquad Nt = \frac{B I_D}{Q V_H} \tag{5.20}$$

where B is the magnetic field, V_H is the Hall voltage, and N is the carrier density. If the thickness is known, the carrier density and therefore the net doping level are known.

In SOI starting material, the Si film is unintentionally doped and fully depleted, so that no current would flow in response to the applied voltage. However, a voltage applied to the substrate can be used to create an inversion or accumulation layer in the same way that the device can be used as a pseudo-FET, and the sheet carrier density and Hall mobility are those of inversion or accumulation carriers. The device of FIGURE 5.34 has the advantage that it can be used both as a pseudo-FET, where one end contact is the source and the other the drain, and as a Hall effect device. EQNS (5.19) and (5.20) apply to these FET-created layers as well as to uniformly doped material, but the carrier density N cannot be obtained by dividing by a thickness in this case. It is the sheet carrier density Nt that is obtained.

The Hall mobility and field effect mobility are not the same because different scattering mechanisms are involved. The field effect mobility as measured by the pseudo-FET is dominated by surface and interface scattering, while the Hall mobility has the additional factor of a different relaxation time. In bulk material, the two mobilities are related by:

$$\frac{\mu_H}{\mu_F} = \frac{\langle \tau^2 \rangle}{\langle \tau \rangle^2} \qquad (5.21)$$

where τ is the mean time between collisions. For Si at room temperature, this ratio is approximately 1.18, so that the Hall mobility is larger than the FET mobility by this ratio.

TABLE 5.3 shows the Hall mobility and FET mobility and the corresponding sheet carrier densities measured on the same samples for several SOI wafers. The Hall mobility was measured at several values of gate (substrate) voltage and the mobilities were extrapolated to $V_G - V_{TH} = 0$ in order to compare the Hall mobility with the values of μ_0 and μ_{EFF} determined by EQNS (5.4) and (5.7). The Hall mobility is usually larger than the FET mobility

TABLE 5.3   Mobilities and sheet carrier densities obtained by Hall effect and FET measurements.

Si, BOX thickness, Å	μ_0 (FET)	μ_H (Hall)	Nt, carriers/cm^2	$\mu_H/1.18$
1580, 1489	644	793	1.2×10^{11}–1.3×10^{12}	672
1605, 1403	633	758	2×10^{11}–1.6×10^{12}	642
1557, 1437	624	766	1.5×10^{11}–1.5×10^{12}	650
974, 1783	644	743	4.8×10^{11}–1.4×10^{12}	630
700, 1430	683	850	8.0×10^{10}–9.0×10^{11}	720

by slightly more than the 1.18 ratio for bulk Si, which might imply that the ratio in EQN (5.21) has to be modified for thin Si films on SOI substrates. In any case, the field effect and corrected Hall mobilities are usually within 5% of each other, which is a strong indicator of the validity of both measurement techniques.

5.8.2 Corona discharge

Corona discharge has been used as a non-contact, non-destructive method for characterising bulk Si. It is basically a way of modulating the surface potential of a substrate without an electrode. In this method, an ionic charge is placed on the surface of an oxidised sample by putting it in the vicinity of a high voltage source which causes breakdown of air molecules. Ions are then accelerated toward the wafer and accumulate on the surface. The high charge density is equivalent to a gate voltage, and the surface potential of the Si is modified in accordance with the polarity and density of the charge. A Kelvin probe is used to detect and measure the surface potential, and the technique can be used to determine flatband voltage, oxide thickness, interface state density, doping level and lifetime [5, 28]. The device has been called a COS, for corona oxide semiconductor, in analogy with MOS structures.

One of the most useful applications for corona discharge is in measuring oxide quality. The ion density deposited on the oxide surface can only disappear by either neutralisation from the ambient or leakage through the oxide. Since free ion densities in the room air are likely to be low, most of the discharge occurs through leakage, and leakage occurs first at weak points in the oxide. The Kelvin probe can measure the charge density, and a wafer scan of the charge density is a map of the weak spots in the oxide [29]. This way of studying oxide integrity is applicable to SOI as well as bulk Si. Defect maps of oxide integrity have been made for 20 nm oxides on SIMOX wafers with 200 nm Si films and 150 nm buried oxides [30]. However, defect maps of thinner oxides useful for sub-quarter micron technology nodes have not been reported.

It might be thought that corona charges placed directly on the Si layer of SOI material could be used to study defects in the Si film, since the Si film is separated from ground by the buried oxide. Most Si layers, however, will not hold a deposited charge for more than a few minutes [30]. Any conducting defect in the buried oxide, such as a pinhole or other weak region, causes the entire Si film to be quickly discharged, even on a 200 mm wafer. This requires an oxide on the SOI surface just as in bulk material. COS devices have not been reported on SOI for studying lifetimes, interface states, etc., for the same reason that photovoltage is not used for SOI films: the Si layers are much thinner than the depletion regions needed for such measurements. However, COS measurements can be very useful for gate oxide integrity studies.

5.9   CONCLUSION

This chapter has outlined some of the techniques used to study the quality of SOI/SIMOX starting material, including reflectometry, ellipsometry, X-ray reflection, pseudo-FETs, MOS capacitors, BOX capacitors, gate oxide capacitors, recombination and generation lifetime, Hall measurements and contactless corona discharge. All of these techniques have their uses, advantages and disadvantages. The electrical measurements made on unprocessed material are not directly applicable to material in finished circuits and devices which usually involve doping levels 2 to 3 orders of magnitude higher and many varieties of process steps, but they are a good indication of the quality of starting material and are much cheaper and faster to implement. Of all the techniques, spectroscopic ellipsometry and HgFETs have proven the most useful for both experimental material in development and production material quality control: ellipsometry because it supplies the basic thickness and composition measurements with high accuracy and the Hg-based pseudo-FET because it yields so many important material parameters easily and quickly. As material advances are made, especially material with very thin Si layers, SiGe, and strained Si for advanced technology nodes, new characterisation techniques may have to be developed. However, the workhorses of spectroscopic ellipsometry and pseudo-FETs, particularly in the HallFET configuration, are applicable down to sub-10 nm thicknesses and can be expected to play a valuable role in continued material development and quality control.

REFERENCES

[1]   G.G. Shahidi [*IBM J. Res. Dev. (USA)* vol.46 (2002) p.121–31]

[2]   Semiconductor Industry Association, Nat. Tech. Roadmap for Semiconductors, 2003 edition, http://public.itrs.net

[3]   H.J. Hovel [*1996 IEEE Int. SOI Conf. Proc.* (Oct. 1996) p.1; *Future Fab. Int.* vol.1 no.2 (1997) p.225]

[4]   S. Cristoloveanu, S.S. Li [*Electrical Characterization of Silicon-on-Insulator Materials and Devices* (Kluwer Academic, 1995)]

[5]   D.K. Schroder [*Semiconductor Material and Device Characterization* (Wiley-Interscience, 1998)]

[6]   H.-S. Wong [*IBM J. Res. Dev. (USA)* vol.46 (2002) p.133–68]

[7]   G. Wohl et al. [*Thin Solid Films (Switzerland)* vol.369 (2000) p.175–81]

[8]   N. Sugiyama et al. [*Thin Solid Films (Switzerland)* vol.369 (2000) p.199–202]

[9]   S. Cristoloveanu, D. Munteanu, S.T. Liu [*IEEE Trans. Electron Devices (USA)* vol.47 (2000) p.1018–26]

[10] H.J. Hovel [*1997 IEEE Int. SOI Conf. Proc.* (Oct. 1997) p.180–1; *Solid State Electron. (UK)* vol.47 (2003) p.1311–33]

[11] X. Zhao, D. Ioannou [*1999 IEEE Int. SOI Conf. Proc.* (Oct. 1999) p.52–3]

[12] L. Jastrzebski, G. Cullen, S. Soydan [*J. Electrochem. Soc. (USA)* vol.137 (1990) p.303–5]

[13] J. Linnros [*J. Appl. Phys. (USA)* vol.84 (1998) p.275–83]

[14] G.S. Kousik, Z.G. Ling, P.K. Ajmera [*J. Appl. Phys. (USA)* vol.72 (1992) p.141–6]

[15] E. Yablonovitch et al. [*Phys. Rev. Lett. (USA)* vol.57 (1986) p.249–52]

[16] Y. Hayamizu et al. [*J. Appl. Phys. (USA)* vol.69 (1991) p.3077–81]

[17] P. Ronsheim [private communication]

[18] P.-C. Yang, S.S. Li [*Appl. Phys. Lett. (USA)* vol.61 (1992) p.1408–10]

[19] Y.-S. Chang, S.S. Li, P.-C. Yang [*Solid-State Electron. (UK)* vol.38 (1995) p.1359–66]

[20] P-C. Yang, S.S. Li [*Solid-State Electron. (UK)* vol.35 (1992) p.927–32]

[21] S. Ibuka, M. Tajima [*Jpn. J. Appl. Phys. (Japan)* vol.39 (2000) p.L1224–6]

[22] S. Cristoloveanu et al. [*1996 IEEE Int. SOI Conf. Proc.* (Oct. 1996) p.160–1]

[23] D. Ioannou et al. [*IEEE Electron Device Lett. (USA)* vol.11 (1990) p.409–11]

[24] D.P. Vu, J.C. Pfister [*Appl. Phys. Lett. (USA)* vol.47 (1985) p.950–3]

[25] L.J. McDaid et al. [*IEEE Electron Device Lett. (USA)* vol.12 (1991) p.318–20]

[26] H. Seghir et al. [*1991 IEEE Int. SOS/SOI Conf. Proc.* (Oct. 1991)]

[27] T. Ouisse et al. [*IEEE Trans. Electron Devices (USA)* vol.38 (1991) p.1432–44]

[28] D.K. Schroder et al. [*Solid-State Electron. (UK)* vol.42 (1998) p.505–12]

[29] P. Edelman, A. Savchouk, J. Lagowski [*Inst. Phys. Conf. Ser. (UK)* no.160 (1997) p.141–4]

[30] D. Guidotti, H.J. Hovel [unpublished work]

Chapter 6

SIMOX material technology from R&D to advanced products

D.K. Sadana

Silicon-on-insulator (SOI) based devices and circuits increase chip speed, lower voltage operation and enhance resistance to cosmic ray induced "soft error" events. Advanced CMOS logic and memory applications require ultra-thin SOI, with Si layers of less than 1000 Å. The most direct and powerful method to form cost efficient SOI is by separation of implanted oxygen (SIMOX). This method utilises O^+ implantation into a heated Si substrate ($>200°C$) at doses $>2 \times 10^{17}\,cm^{-2}$ followed by high temperature annealing (above 1300°C), to form a buried SiO_2 layer. Recent advances in improving SIMOX quality by the modified low-dose process (MLD) will be described. It will be shown that the quality of the modern MLD SIMOX is comparable to that of bonded SOI. Functional products including microprocessors, SRAM memories and high frequency RF circuits utilising IBM's $0.18\,\mu m$ and $0.13\,\mu m$ CMOS technologies show equivalent yield on both bonded SOI and MLD SIMOX.

6.1   INTRODUCTION

Silicon-on-insulator (SOI) devices were first developed for early satellite and man-in-space exploration systems in the 1960s. The main advantage of SOI devices was their resistance to ionisation from solar wind radiation and voltage isolation of the chips. Most of the early SOI devices were made with silicon-on-sapphire (SOS) wafers. Replacement of SOS by SOI was accomplished by the SIMOX process (separation by implantation of oxygen) in the 1970s [1]. Today, SOI wafers are produced mainly by two methods, namely, wafer bonding [2] and SIMOX.

Material technology to mass produce both bonded SOI and SIMOX wafers is maturing rapidly. Production capacity for 200 mm and smaller diameter wafers is already in place to meet the world's SOI wafer demand. Recently, small scale production

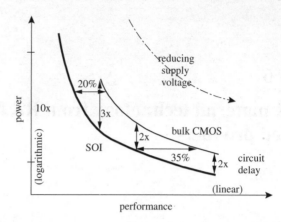

FIGURE 6.1   Power–performance curves for bulk and SOI CMOS technologies.

of 300 mm SOI wafers by both methods has been accomplished. Furthermore, increased production capacity of 300 mm SOI wafers is already being put into place to meet the anticipated demand of the silicon IC industry. Manufacturability of CMOS logic circuits is transparent to both bonded or SIMOX wafers. Similarly, power-performance enhancement is transparent to both types of substrate. Compared to similar circuits on bulk Si, SOI CMOS can run at 20% to 50% higher switching speeds and with 2 to 3 times lower power requirements (FIGURE 6.1). These improvements in speed and power usage for SOI CMOS are equivalent to accelerating 1 to 2 generations of transistor scaling on bulk Si [3]. In this chapter we will focus exclusively on SIMOX material technology and discuss its applications to mainstream CMOS memory and logic products.

6.2   PRODUCTS AND SCALABILITY

SIMOX technology has evolved from the standard dose process of the 1970s which required oxygen doses of $>1.5 \times 10^{18}$ cm^{-2} [1, 4] to state-of-the-art modified low-dose process (MLD SIMOX) which requires oxygen doses in the low 10^{17} cm^{-2} range [5]. In the 1980s and 1990s, however, a variety of SIMOX processes were developed including low-dose and medium dose processes [6–8]. FIGURE 6.2 summarises the evolution of various SIMOX processes aiming at various BOX and SOI thicknesses. Most modern versions of SIMOX have excellent thickness and thickness uniformity control, high integrity of BOX with very low short density (<1 cm^{-2}), and low levels of metallic contamination ($<10^{11}$ cm^{-2}). There are many reasons why SIMOX is an

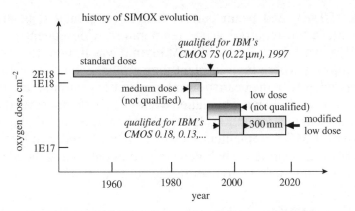

FIGURE 6.2   A historical perspective of advancements in SIMOX and its adoption in CMOS logic technologies.

attractive process for SOI fabrication. First, it is quite manufacturable as it involves only two main process steps, implantation and annealing, both of which have been practised by the Si IC industry for many years. Second, it is a very scalable process in terms of both implantation dose and beam current. There is a straightforward relationship between the implant dose and cost, i.e. the lower the implantation dose, the lower the SIMOX cost. Alternatively, the higher the beam current for a given dose, the lower the cost assuming the equipment cost does not increase in proportion to the beam current. There are challenges, however. With lowering of the dose, especially when the peak concentration is well below $(<\frac{1}{2})$ that required for stoichiometric SiO_2, a discontinuous BOX forms. New implant and/or annealing concepts have to be applied to enhance coalescence of oxide precipitates and formation of a continuous BOX. This will be discussed in the ensuing sections.

6.3   HIGH TEMPERATURE OXYGEN IMPLANTATION

Since the SIMOX process requires very high fluences of oxygen ranging from low-mid 10^{17} O^+ cm^{-2} for low-dose versions to $>1.5 \times 10^{18}$ cm^{-2} for the high dose version (standard dose), the Si substrate temperature is raised to a few hundred degrees (typically above 500°C) to enhance dynamic annealing of implantation induced damage [9, 10]. In earlier generations of commercial oxygen implanters high power generated by high ion beam currents (40–60 mA) was used to achieve wafer temperatures >500°C. However, this approach greatly limited the range of implant energies and beam currents under which the SIMOX process could be practised. For example, beam energies had to be greater than 150 keV and beam current had to be greater than 40 mA to achieve the required wafer temperature. Consequently, early SIMOX work in the sub 100 keV regime was limited to research laboratories where high implant temperatures could be achieved independently of the beam power by using conventional resistively heated wafer holders. This latter work clearly demonstrated dose regimes where a continuous BOX could be formed at low energies. Modern SIMOX implanters, however, provide independent control of wafer heating via halogen lamps thus enhancing the SIMOX process window down to energies lower than 100 keV [11].

6.3.1 SIMOX: as-implanted microstructure vs oxygen dose at >500°C

The microstructure of as-implanted SIMOX depends on the implant conditions. For example, for standard dose SIMOX where the dose is $\sim(1.8–2.0) \times 10^{18}$ cm^{-2} at 200 keV, a buried oxide layer has already formed in the implanted region (FIGURE 6.3(a)). Highly damaged and irregular structure is present on either side of this buried oxide layer. On the other hand, in the case of low-dose SIMOX with a dose of $<4 \times 10^{17}$ cm^{-2} at >150 keV only a buried band of damage clusters intermixed with oxide precipitates is present in the implanted region. The density of SiO_x precipitates submerged in the damage band increases with dose until a near-continuous but thin oxide region forms at a dose of $(5–6) \times 10^{17}$ cm^{-2} (FIGURE 6.3(b)). The thickness as well as continuity of the buried oxide region increases as oxygen dose in the as-implanted SIMOX increases. There appears to be a correlation between increased oxide precipitation in the damage band in the as-implanted material and an increase in the density of Si inclusions in the annealed wafers [12]. The Si inclusion density begins to decline at doses which are significantly higher than those theoretically needed to achieve stoichiometric SiO_2. Excess implanted oxygen consumes the disordered Si and SiO_x precipitates or other nucleation sites by internal oxidation [13].

(a) as-implanted

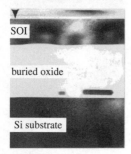

(b) annealed

⊢ - - - - - - - ⊣
0.4 μm

FIGURE 6.3   Cross-section TEM micrographs of standard dose SIMOX before and after anneal. Note the formation of a buried oxide layer in the as-implanted sample (a). The surface region in (a) is crystalline albeit highly defective.

6.4   SIMOX: ANNEALED (MATERIAL)

6.4.1   Standard dose

Annealing kinetics of standard dose SIMOX are vastly different from those of low-dose SIMOX because of their vastly different oxygen concentrations and damage structures in the implanted region (FIGURE 6.3(a) and 6.4(a)–(e)). Typical annealing for standard dose SIMOX is carried out at temperatures $>1300°C$

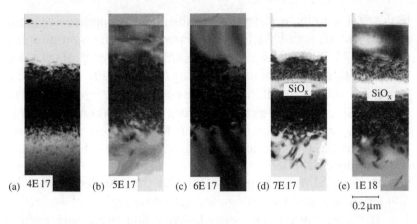

FIGURE 6.4   Cross-section TEM micrographs showing damge distribution in as-implanted SIMOX as a function of the oxygen dose. Note the formation of an oxide layer at doses $>6 \times 10^{17}\,\text{cm}^{-2}$. The surface region in all samples is crystalline due to dynamic annealing during the implant.

for 4–8 hours in Ar or N_2 ambient with low concentrations of O_2 (0.5–2%) [1, 4, 14]. A complex mechanism involving first the growth of oxide precipitates above the BOX followed by their dissolution takes place during annealing. The oxygen released by the oxide precipitates is absorbed by the upper Si/BOX interface increasing the overall thickness of the BOX. Simultaneously, annihilation of voids and other defects at the surface takes place via interaction with Si interstitials, vacancies and their complexes. The end result is an SOI material with a dislocation density in the range $10^5–10^7\,\text{cm}^{-2}$ with an SOI thickness of $<2000\,\text{Å}$, and BOX thickness of $\sim4000\,\text{Å}$. The density of dislocations depends on the implant dose, temperature, and the resultant stresses in the as-implanted material.

6.4.2   Low-dose

Early development of SIMOX from the 1970s to the early 1990s was dominated by the standard dose process. The dose window was defined by the stoichiometry requirements of SiO_2 in Si. Despite its initial success in demonstrating high yielding ICs, the process faces challenges due to its low throughput and high cost. A modern 200 mm SIMOX implanter with a nominal 50 mA of O^+ beam current has the maximum daily output of only ~25 wafers of standard dose SIMOX which corresponds to ~1 wafer/hr [11]. A focused effort has therefore been under way since the early 1990s to develop a high throughput, low-dose SIMOX process for the Si IC industry. The minimum BOX thickness requirement is that which avoids capacitive coupling of devices and circuits formed in the SOI with the underlying Si substrate. Comprehensive studies performed under DARPA's Low Power Electronics programme during 1995–2000 showed that BOX thicknesses of $\geq 1000\,\text{Å}$ are sufficient to avoid capacitive coupling. SOI circuits formed on a BOX of $\sim 1000\,\text{Å}$ maintain the same performance advantage over the bulk-Si technology as that achieved with a 4000 Å BOX. This BOX thickness reduction corresponds to a decrease of four times or more in the oxygen dose compared to that used to form the standard dose SIMOX.

Low-dose SIMOX implant efforts in the last decade can be divided mainly into two categories: (i) a dose regime of 4×10^{17}–$1 \times 10^{18}\,\text{cm}^{-2}$ at energies of $>150\,\text{keV}$ and (ii) a dose regime of $<3 \times 10^{17}\,\text{cm}^{-2}$ at energies of $<100\,\text{keV}$. Similarly, annealing of the implanted substrates can be divided into two categories: (i) non-ITOX (internal thermal oxidation) and (ii) ITOX based. The final product is aimed to create SOI thicknesses of $>1000\,\text{Å}$ for partially-depleted devices, and $<500\,\text{Å}$ for fully-depleted devices. Most of the work, however, has concentrated around processes which create SOI for partially-depleted devices. The learning above can be extended to fully-depleted device applications by either forming a thin SOI layer by increasing oxidation during low-dose SIMOX fabrication, or simply thinning the SOI layer of a pre-formed SOI wafer by thermal oxidation. Alternatively, the SIMOX process in category (ii) can be extended to include partially-depleted device applications by growing an epitaxial Si layer on the finished substrate. The latter may be undesirable for two reasons: (a) additional tooling and process cost, and (b) reduced thickness control of the SOI layer as the epi-Si layer may introduce thickness non-uniformity of 1–2%.

When standard dose annealing conditions are applied to low-dose SIMOX, discontinuous oxide regions centred on peaks of the damage and implanted oxygen profiles form. Development of a manufacturable process which provides a continuous BOX at the lowest possible oxygen dose and with the highest electrical integrity has been the focus of SIMOX studies for the last 10–12 years.

6.4.2.1 Non-ITOX

In this process annealing is conducted with relatively low oxygen levels ($<10\%$). Wafers are typically implanted in the dose range $(4–10) \times 10^{17}\,\text{cm}^{-2}$ in either single or multiple steps followed by annealing at $>1300°\text{C}$ in single or multiple steps. The thickness of the BOX corresponds closely to that expected from the implanted dose. For example, a dose of $4 \times 10^{17}\,\text{cm}^{-2}$ corresponds to a BOX thickness of ~1000 Å (see FIGURE 6.8(b)) [9]. Even though a continuous and structurally "good-looking" oxide can be created by this method, the electrical quality of the BOX, as determined by its breakdown field, is much inferior compared to that of the thermal oxide. Silicon islands embedded in the BOX (FIGURE 6.5) are responsible for deteriorated BOX breakdown properties. The density of these precipitates can vary from 10^5 to $10^8\,cm^{-2}$ depending on the actual dose of oxygen and the anneal conditions employed.

6.4.2.2 ITOX

This annealing process is very effective in improving the BOX quality when the implanted oxygen dose is around $4 \times 10^{17}\,cm^{-2}$ (so called Izumi window). The annealing is conducted at >1300°C in an ambient that consists of inert gas (typically Ar) mixed with high concentrations of oxygen, typically in the range 30–60% [6–8]. Under such annealing conditions, not only does oxygen react with the Si surface to form the surface oxide but it also diffuses into the Si in significant amounts. The upper-Si/BOX interface acts as a sink for the diffusing oxygen and an additional thermal oxide-like BOX is formed on top of the implanted BOX. FIGURE 6.6 shows the relationship between the ITOX induced oxide growth and the thermal oxide growth on the surface during annealing at 1350°C. The electrical quality of this BOX (as determined by its breakdown field) is superior to that of the BOX formed without the ITOX (see FIGURE 6.9). The BOX thickness can be increased in a predictable manner for a given anneal temperature and oxygen concentration. The BOX thickness therefore has two components, the implanted dose component which forms the lower part of the BOX, and the thermal oxide component which forms the upper part of the BOX. FIGURE 6.7 shows the relationship between the ITOX induced oxide and the reciprocal of annealing temperature [5–7]. Typically, ITOX induced BOX thickness is 8–10% of the surface oxide. Under practical implant conditions, however, ITOX induced BOX thickness is limited to <900 Å for the previous generation of implanters which run at a maximum beam energy of 210 keV. The maximum Si thickness that is available to form the surface oxide for a 210 keV O$^+$ implant is limited to ~4000 Å which limits the maximum surface oxide growth to <8500 Å, and hence the maximum ITOX thickness to < 900 Å. FIGURE 6.8(b) shows ITOX data from 180 keV oxygen implants where the maximum Si thickness available is ~2500 Å, and the maximum ITOX is limited to <600 Å. XTEM micrographs of FIGURES 6.8(b) and 6.8(c) compare the BOX in non-ITOX and

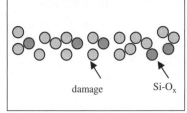

as-implanted

damage   Si-O$_x$

(a)

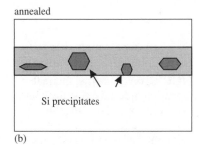

annealed

Si precipitates

(b)

FIGURE 6.5   Schematics of as-implanted damage structure and corresponding BOX structure in low-dose SIMOX prepared by non-ITOX based anneal.

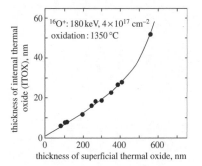

FIGURE 6.6   Internal thermal oxidation induced buried oxide thickness versus the oxide grown on the surface of the SIMOX during annealing at 1350°C [6–8].

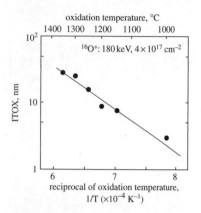

FIGURE 6.7   Internal thermal oxidation induced oxide thickness versus reciprocal of annealing temperature (1/T). The surface oxide thickness was ~400 nm and was kept constant for all temperatures [6–8].

ITOX wafers, and confirm the oxide growth data in FIGURES 6.6 and 6.7. Since the top region of the BOX in ITOX wafers is thermal oxide, the BOX properties, such as breakdown voltage and short density, show marked improvement as shown in FIGURE 6.9 [15].

6.4.2.3   Modified low-dose (MLD) SIMOX

This is a powerful SIMOX manufacturing process which allows the formation of high quality SIMOX over a continuous range of oxygen doses spanning from low-dose ($\sim 6 \times 10^{17}$ cm^{-2}) to ultra-low-dose (1.5×10^{17} cm^{-2}) [5]. The BOX thickness can be varied from 700 Å to 2500 Å whereas the SOI thickness can be varied from 2000 Å to 50 Å depending on the application. The process utilises the following three major steps: (i) implanting a base dose of ^{16}O$^+$ ions (in the range described above) into a hot (>200°C) Si substrate at energies of >100 keV, (ii) cooling down the substrate to nominal room temperature and amorphising a part of the implanted region in (i) by another very low-dose O$^+$implant (the touch up RT implant) [16], and (iii) subsequently annealing the composite structure at >1300°C under conditions which enhance internal thermal oxidation at the implanted region [5]. FIGURE 6.10 shows a schematic of structures in the as-implanted and annealed MLD SIMOX. As in the ITOX process, thermal oxide grows at the upper BOX interface in MLD SIMOX. However, the damage created by the touch up RT implant has two advantages over the non-ITOX and ITOX processes described earlier: (i) it enhances BOX continuity in the dose regime of $<4 \times 10^{17}$ cm^{-2}, i.e. it extends the Izumi window into a much lower dose regime, and (ii) it also enhances ITOX by almost 1.5 times or more compared to the process described in Section 6.4.2.2. Depending on the requirements of the final SOI thickness, values of the

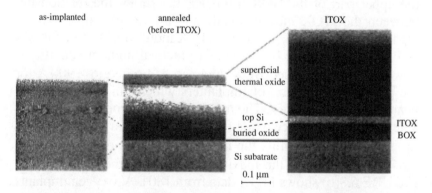

FIGURE 6.8   Cross-sectional TEM micrographs showing the BOX formation with and without ITOX [8]. Implant conditions: 180 keV, 4×10^{17} cm^{-2}. Anneal temp.: 1350°C.

ITOX induced BOX can be 1000 Å or greater (for Si, <500 Å (FIGURE 6.11)). Consequently, the BOX layer can be predominantly thermal-oxide-like. Typical intrinsic breakdown field in state-of-the-art MLD SIMOX (550 Å SOI or below) is ~8 MV/cm which compares quite favourably with the breakdown field of ~9–10 MV/cm in bonded SOI where the BOX is made up of pure thermal oxide (see FIGURE 6.13). The BOX short density in MLD 550 Å is also comparable to that in the bonded SOI (<0.2 cm^{-2}). FIGURE 6.12 shows an XTEM micrograph of the 550 Å MLD SIMOX. This material is presently being qualified for 0.1 μm generation CMOS technology at IBM.

FIGURE 6.13 highlights the scalability of the SIMOX process as SOI applications move towards 0.1 μm and beyond CMOS technology generations. The intrinsic electrical quality of the BOX improves dramatically as the base dose of O$^+$ decreases. This phenomenon occurs because the ITOX fraction of the BOX increases with respect to the implanted fraction of the BOX as the O$^+$ dose.

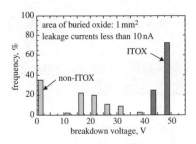

FIGURE 6.9   Improvement in the electrical quality of the BOX by ITOX. Both the breakdown field and short density improve with ITOX [15].

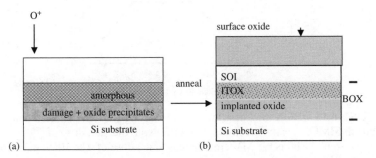

FIGURE 6.10   Schematics of modified low-dose process: (a) as-implanted structure, (b) annealed structure.

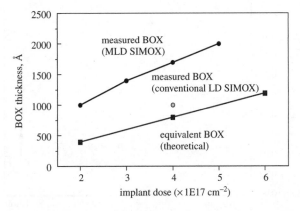

FIGURE 6.11   Modified low-dose SIMOX showing enhanced ITOX (see calculated versus experimental curves), formation of a continuous BOX at doses to ~2E17 cm^{-2} (well below the Izumi dose window), and a continuum of doses where a continuous BOX can be formed [5].

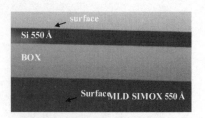

FIGURE 6.12   An XTEM micrograph showing state-of-the-art MLD 550 Å SIMOX (based on a patent in [5]).

From a manufacturing point of view, one produces better SIMOX at a lower cost! A throughput of >6 wafers/hr is achievable in production mode with the current generation of oxygen implanters if O^+ doses of $<3 \times 10^{17}$ cm^{-2} are used. The structural and electrical qualities of the BOX in MLD SIMOX are expected to continue improving concomitant with decreasing oxygen dose. This trend is driven by the SOI thickness scaling for future generations of CMOS technology. Thinner SOI translates into ITOX as more Si is available for oxidation (so long as the O^+ implant energy is fixed), and therefore for a fixed BOX thickness less oxygen dose is required. FIGURE 6.14 shows the SOI thickness roadmap for various CMOS technologies.

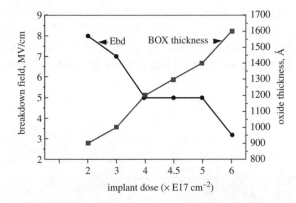

FIGURE 6.13   BOX thickness and corresponding BOX breakdown field (Ebd) in SIMOX versus the implanted dose. Note that at a dose of 2×10^{17} cm^{-2} Ebd is ~ 8 MV/cm whereas at a a dose of 4×10^{17} cm^{-2} Ebd drops to 5 MV/cm.

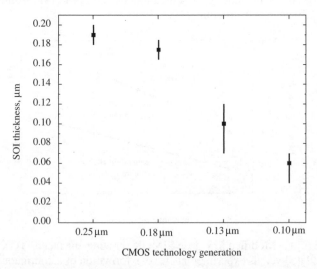

FIGURE 6.14   SOI thickness versus CMOS technology generation for partially-depleted device design.

6.4.2.4   *Standardisation of SIMOX via MLD SIMOX*

There has been a deliberate move in the last few years to standardise SIMOX processes for high volume IC production. Since SIMOX can be produced in a variety of ways, it has always been difficult in the past to decide which choice of SIMOX is most suited for IC applications. In addition, there was no compelling reason to try out product based test vehicles, such as microprocessors, high density SRAMs etc. that could be tried to determine which particular make of SIMOX was most qualified for high volume IC requirements. However, since 1997 when IBM announced its CMOS SOI technology, it has been shown that microprocessors built on MLD SIMOX yield equivalent to those built on bulk Si. Now, more than three generations of CMOS SOI technology have been qualified on MLD SIMOX, and this material is becoming one of the leading candidates for SOI wafer supply for the semiconductor industry.

6.5   CHARACTERISATION OF SOI MATERIALS

Not only has the task of fabrication of SOI material been challenging but equally challenging has been its screening for product worthiness. Many of the optical and electrical characterisation techniques developed earlier for bulk Si cannot be directly applied to SOI materials because of the presence of the BOX. Furthermore, characteristics of the defects that limit device and circuit yield in SOI are still not fully understood. Despite these limitations, remarkable progress has been made in evaluating SOI to a point where one can sort the good material from the bad material with high confidence. Some of the main parameters for evaluation include: (i) SOI thickness and thickness uniformity, (ii) BOX thickness and uniformity, (iii) HF-defect density, (iv) dislocation density, (v) surface roughness, (vi) surface pitting, (vii) concentration of metallic contaminants, (viii) carrier mobilities in the SOI region, and (ix) fixed charge in the BOX, and at the Si/BOX interface. TABLE 6.1 summarises typical physical and electrical parameters of MLD SIMOX and compares them with those of bonded SOI [14].

6.5.1   Physical characterisation

6.5.1.1   *Thickness control and uniformity*

SOI thickness and thickness uniformity are two key parameters that can impact the device and/or circuit operation and reliability. These become even more critical for circuits based on fully depleted device design. Thickness and thickness uniformity are

TABLE 6.1   Physical and electrical properties of modern
commercial SIMOX.

Parameter	Unit	Bonded SOI	MLD SIMOX
SOI thickness uniformity (1 sigma)	Å	10–15 (−)	6–9 (+)
HF-defects	cm^{-2}	<0.5	<0.5
Secco etch pits	cm^{-2}	400–700	400–700
Surface roughness (10 μm × 10 μm)	Å	1.0–2.0 (+)	4–7 (−)
Electron mobility	cm^{-2}/V s	500–600 (−)	550–650 (+)
Hole mobility	cm^{-2}/V s	200–230	200–230
BOX short density	cm^{-2}	<0.5	<0.5
BOX breakdown field	MV/cm	>8 (+)	7–8 (−)
BOX charge	cm^{-2}	<3.0 × 10^{11}	3.0 × 10^{11}
Interface charge	cm^{-2}	<3.0 × 10^{11}	3.0 × 10^{11}

generally measured by spectroscopic ellipsometry, or by other multi-wavelength spectroscopic tools. Thickness uniformity of <1% is routinely achieved for SIMOX. Highly sophisticated thickness mapping tools with fast turnaround time have been developed, such as IPEC's Acumap 2. In this tool, over 30,000 measurements are made in about 1 minute and the data are converted into printable thickness maps within 2 minutes. Examples of typical thickness maps of SOI layers are shown in FIGURES 6.15(a) and 6.15(b) for commercial bonded SOI and low-dose SIMOX, respectively, for 0.18 μm CMOS technology. It is clear that highly uniform SOI layers can be produced by both methods. FIGURES 6.16(a) and 6.16(b) show that this thickness uniformity can be maintained on state-of-the-art 300 mm SIMOX and bonded SOI. Both technologies have been able to demonstrate highly uniform SOI thicknesses with the uniformity range of <40 Å across the entire wafer for an SOI thickness of ~550 Å.

6.5.1.2   HF defects

One of the simple but powerful techniques for SOI evaluation is HF etching. Starting SIMOX or any other kind of SOI material

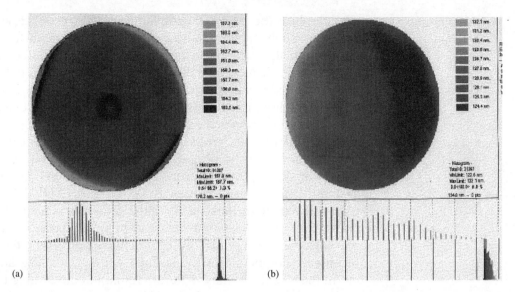

FIGURE 6.15 (a) A typical thickness map for commercial bonded SOI prepared by the Unibond process. SOI and BOX thicknesses are ~1550 Å and 1450 Å, respectively. Max-min thickness variation is ~100 Å. (Courtesy H. Hovel, IBM.) (b) A typical thickness map for low-dose SIMOX. SOI and BOX thicknesses are 1250 Å and 1450 Å, respectively. Max-min thickness variation is ~80 Å. (Courtesy H. Hovel, IBM.)

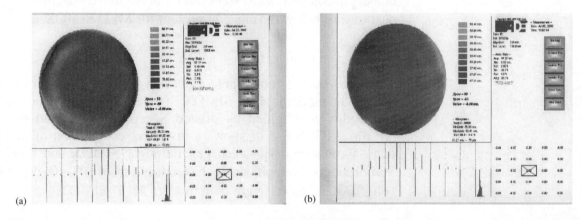

FIGURE 6.16 (a) A typical thickness map for commercial bonded SOI prepared by the Unibond process. SOI and BOX thicknesses are ~550 Å and 1450 Å, respectively. Max-min thickness variation is ~40 Å. (Courtesy, H. Hovel, IBM.) (b) A typical thickness map for low-dose SIMOX. SOI and BOX thicknesses are 550 Å and 1400 Å, respectively. Max-min thickness variation is ~30 Å. (Courtesy, H. Hovel, IBM.)

is dipped in concentrated HF (49%) for several minutes (typically 20–30 min) to highlight any weaknesses in the SOI layer. If there is any path for HF to diffuse through the SOI layer into the underlying BOX, an HF defect is created [17, 18]. An HF defect is disc-shaped and its size depends on the length of the HF exposure, and type of the defect being etched. HF defects in SIMOX can originate for a number of reasons; however, particles blocking oxygen implantation is by far the most dominant mechanism for HF defects. If the particle shifts the peak of the implanted oxygen profile towards the Si surface such that the resulting BOX intersects the surface, an HF defect is created. There may be other sources of HF defects as well. For example, if there is a catalytic reaction between the particle and implanted Si surface such that an opening in the surface is created during high temperature annealing, an HF defect can result. Metal particles, even in the absence of implant blocking, can be the source of HF defects. Metal can react with the Si surface during the hot implant to form a silicide or a silicate region. HF reacts with these silicided regions to form an opening in the SOI allowing HF to find its way to the underlying BOX. The size of HF defect in this case will depend on the tenacity of the silicide or silicate. Typically there is a finite incubation period (which can extend from a few seconds to a few minutes) during which HF reacts with the silicide or silicate compound at the surface without creating an undercut in the BOX, and may therefore remain unobservable under an optical microscope. A prolonged HF exposure (>20 min) is therefore recommended to highlight such tenacious defects with a long incubation time. Metallic induced HF defects have a high probability of impacting the device and circuit yield adversely. HF-defect density in commercial SIMOX has reduced from 5–10 cm^{-2} in the past to well below 0.5 cm^{-2} at present for MLD SIMOX.

6.5.1.3   Secco defects

There has been a remarkable reduction in threading dislocation density in modern MLD SIMOX. These threading dislocations are readily highlighted by diluted Secco etch followed by an HF dip. FIGURE 6.17 shows a historical comparison of defect density reduction with each generation of SIMOX and bonded SOI products. Etch pit density has reduced from 10^6–10^7 cm^{-2} in "old SIMOX" to $<10^3$ cm^{-2} in modern MLD SIMOX.

6.5.1.4   Surface roughness

Surface roughness and pitting can routinely be measured by atomic force microscopy (AFM). Typical RMS values of surface roughness are 7–10 Å on a 10 μm × 10 μm area for standard dose SIMOX. The roughness for MLD SIMOX is typically 3–7 Å, which is nearly half the standard SIMOX surface roughness value. It is generally found that if AFM is performed on the BOX after removing the top SOI layer, the roughness depends on the annealing procedure used. Typically, there is a high frequency roughness when a non-ITOX anneal is used. With an ITOX anneal, however, a much smoother BOX is created with RMS values close to that of the upper SOI layer.

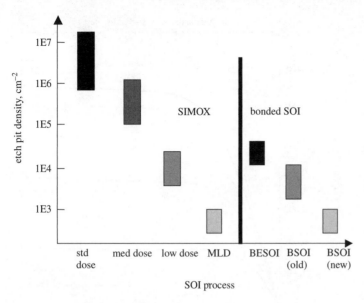

FIGURE 6.17   A historical perspective of crystallographic defect reduction in SIMOX and bonded SOI materials.

6.5.2   Electrical characterisation

6.5.2.1   HgFET: SOI

The presence of the BOX in SOI material makes it possible to form a pseudo-MOSFET by using BOX and substrate as the gate oxide and gate electrode, respectively, and two contacts on the SOI layer as source and drain. Earlier pseudo-MOSFET work utilised two point contacts on the SOI region for source and drain connections [19, 20]. However, the point contacts act as Schottky barriers and the transfer characteristics of the pseudo-FET are pressure sensitive. More recently, an HgFET technique has been developed in which a combination of broad area Hg electrodes coupled with special surface treatment with $HF:H_2O$ is used to overcome the limitations of point contacts [21]. Measurements in the linear region show that the Hg electrodes are Ohmic to electrons and Schottky-like to holes immediately after the surface treatment, but become Ohmic to holes and Schottky-like to electrons after a certain period of time. Therefore, both NFET and PFET transfer characteristics can be obtained by the HgFET technique, which was not possible by the point contact pseudo-FET technique. FIGURES 6.18 and 6.19 include Id–Vd curves for state-of-the-art commercial SIMOX and bonded SOI wafers, respectively. Both figures show very comparable electrical quality of the SOI layer. It is interesting to note that contrary to common belief, low-field electron and hole mobilities of MLD SIMOX are quite close in values to those of the starting Si substrate (TABLE 6.1). Fixed charges

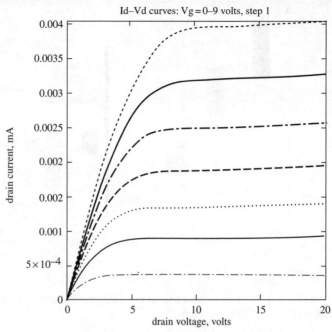

Id–Vd curves: Vg = 0–9 volts, step 1

Hg FET Id–Vd curves: SIMOX; Si = 1485 A, BOX=1473 A

FIGURE 6.18   Id–Vd curves from MLD SIMOX by the HgFET technique. It is clear that the SOI layer has high electrical quality. (Courtesy H. Hovel, IBM.)

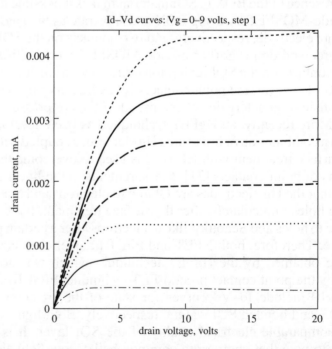

FIGURE 6.19   Id–Vd curves from commercial bonded SOI by the HgFET technique. It is clear that the SOI layer has high electrical quality. (Courtesy H. Hovel, IBM.)

at the upper Si/BOX interface and in the BOX are typically quite low. The BOX integrity as measured by its breakdown field and number of shorts (or low breakdown field sites) is extremely good for both bonded SOI and MLD SIMOX ($<0.2\,\text{cm}^{-2}$).

6.5.2.2 *BOX: shorts and intrinsic properties*

BOX shorts in SIMOX are primarily created by the particles which deposit on the surface during implantation and block the O^+ beam from entering the Si surface. The BOX under the particles is thinner than its surrounding, discontinuous or altogether absent depending on the size of the particles. Until recently, typical BOX short density in low-dose SIMOX was $1-2\,\text{cm}^{-2}$. This short density has been reduced by an order of magnitude in modern MLD SIMOX by reducing particle generation in the implanter, and by adopting improved cleaning procedures to remove particles. The intrinsic property of the BOX in SIMOX is dominated by its stoichiometry. Typically, BOX formed in SIMOX has excess Si. The density of excess Si increases with increasing oxygen dose in the implanted region. In fact, Si precipitates of $>200\,\text{Å}$ can be seen at doses of $>4 \times 10^{17}\,\text{cm}^{-2}$ at energies of $170-200\,\text{keV}$. Such a BOX has a poor breakdown field ($<5\,\text{MV/cm}$) and is dominated by a huge number of mini breakdowns at much lower fields ($2-3\,\text{MV/cm}$). The solution needed to improve BOX intrinsic quality lies in using lower oxygen doses to create the BOX. TABLE 6.1 compares BOX properties of modern MLD SIMOX with bonded SOI [14].

6.6 SIMOX: PRESENT AND FUTURE DEVELOPMENTS

The material quality (as determined by physical and electrical characterisation described above) of SIMOX has improved markedly in the last two decades. This progress has accelerated dramatically in the last few years since SOI product announcements by many major IC manufacturers. There has been a push to standardise SIMOX products using MLD SIMOX based processes. Large diameter MLD SIMOX wafers of up to 300 mm are now commercially available. Early results from these 300 mm wafers show that the thickness uniformity range for a 550 Å SOI layer is $<35\,\text{Å}$ [22, 23].

There have been some noteworthy improvements in SIMOX equipment in the last 2–3 years: (i) O^+ implantation into

300 mm wafers, (ii) O^+ beam currents of up to 100 mA, (iii) O^+ beam energy of up to 240 keV, and (iv) particle levels of <1000 (0.2 μm).

Typically, a majority of particles are generated at pins which hold wafers. Other sources of particles include beam line components that may shower particles when exposed to the beam either due to slight misalignment in its trajectory, or due to high space charge that causes beam blow-up. However, there still remains ample room for improving cost of ownership for the SIMOX process. This could be addressed by building SIMOX implanters that could provide O^+ beam currents of several hundreds of mA, in conjunction with MLD SIMOX processes with much reduced oxygen dose than that in use at present. More importantly, a dramatic reduction in wafer handling time during implantation is required. The handling time in the state-of-the-art implanter is nearly $\frac{1}{3}$ of the total implant time. Processing issues that need attention for 300 mm SIMOX include reduction of slip at the edges of a 300 mm SIMOX wafer, and increased flat zone in the high temperature furnace to increase wafer batch size to reduce cost of ownership. In the long term, the implanter needs to be further modified to (i) increase flexibility of choosing ion species other than O^+, such as O_2^+, N^+, N_2^+, etc., (ii) increase implant temperature capability to higher than 600°C, and (iii) widen the implant energy window in both the low and high energy regimes, e.g. 10–300 keV, rather than the present energy regime of 40–240 keV. These improvements are necessary to allow a wider range of SOI applications for the SIMOX process.

From a SIMOX annealing point of view, the choice between a vertical and a horizontal furnace is still open for discussion. There are pros and cons for both designs. Pros for the horizontal design include large wafer load (>100), and less probability of slip, warp or bow. Cons include a larger footprint for a horizontal furnace, and lack of advanced robot handling. Pros for a vertical furnace include smaller footprint, and robotic handling. Cons include higher probability of slip, warp and bow. There is still room for improvement of SiC hardware for both furnace designs. The wafer support mechanism for 300 mm wafers needs to be quite robust in order to anneal a batch of 50 or more at >1300°C.

6.7 ADVANCED PRODUCTS ON SIMOX: GENERIC AND PATTERNED

Commercialisation of SOI based logic and memory products is actively under way. Many major IC manufacturers have either already developed or are in active pursuit of developing SOI based products. Typical products include programmable gate arrays, static or dynamic memory (SRAM or DRAM), and microprocessors. An example of the most advanced application of SOI is a giga processor test chip developed by IBM as shown in FIGURE 6.20. This test chip is fabricated using IBM's 0.18 μm CMOS technology and contains over 170 million transistors. Functionality at or above 1.3 GHz has been demonstrated in recent tests.

SOI also has enormous potential in extending performance advantage from chip level to system-on-chip level (SOC). It is believed that future performance increases beyond technology enhancements will come from SOC, and patterned SOI will play a significant role for SOC technology. There are several approaches to building SOC on patterned SOI. In one approach logic circuits will be fabricated on the patterned SOI region, whereas DRAM and analogue circuits will be fabricated on an adjacent bulk-Si region [24]. Another approach includes fabrication of cell and sense amplifiers on the bulk Si but peripheral circuits on the SOI. Yet another approach is to fabricate both the memory (DRAM) and logic circuits on SOI. It is too early to predict which of these approaches will be the front runner in future. This will be dictated by several factors which are technology and pattern specific, such as the quality of the SOI (retention time for DRAMs) region, control of lateral straggle, step height at the boundary between the SOI and bulk-Si region, and the presence of dislocations at the SOI/bulk-Si boundary, just to name a few.

FIGURE 6.21 shows schematically how patterned SOI regions are created by the SIMOX method. Major steps include (a) hard

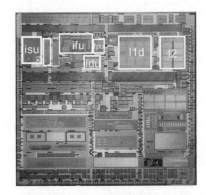

FIGURE 6.20   A schematic layout of IBM's Power4™ processor.

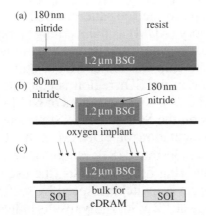

FIGURE 6.21   Formation of patterned SOI regions by the MLD SIMOX process.

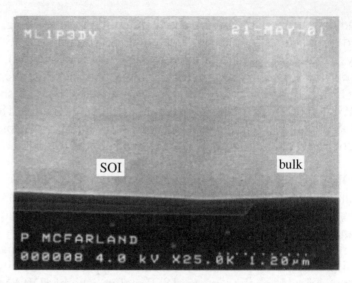

FIGURE 6.22   Reduced defect generation at the mask edge of a patterned SIMOX with the low-dose process [25].

mask film deposition and lithography, (b) 80 nm nitride spacer formation, (c) oxygen implantation (followed by high temperature annealing).

FIGURE 6.22 shows a cross-sectional SEM micrograph of a patterned SOI substrate formed by the MLD SIMOX process. Such structures have been used to demonstrate SOC applications for both 0.18 μm and 0.13 μm eDRAM technologies. In this case, trench-based eDRAM macros were created in bulk Si while high-performance logic circuits were built on SOI. These results show that integrating eDRAM in an SOI-based technology is feasible, and more importantly, paves the way for system-on-chip integration using the SIMOX technology.

A feasibility study for 0.18 μm eDRAM technology at the product level was performed on patterned SOI using 64 Mb DRAM addressable memory (ADM) as the learning vehicle. In this study a checkerboard pattern was used to create patterned SIMOX such that every alternate chip was either SOI or bulk Si. It was shown that retention time in the SOI region is equivalent to that in the bulk Si (FIGURE 6.23). Yield monitors of 64 Mb DRAM showed equivalent yields on the SOI and bulk-Si regions [18]. Recently, successful implementation of eDRAM technology in 0.13 μm technology was demonstrated (8 ns random access, 0.31 μm^2 cell) [25]. Based on the ring oscillator tests, the use of 0.13 μm SOI logic devices improves switching speed by ∼20% over 0.13 μm bulk-Si technology at 1.2 Vdd. Early results indicate that eDRAM yield and retention characteristics are comparable to those of bulk Si with initial retention fails occurring at 128 ms [25]. This

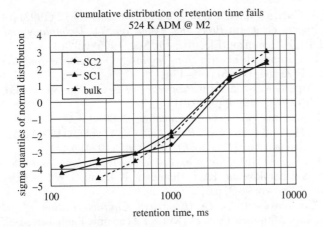

FIGURE 6.23   A comparison of retention times in bulk Si, annealed bulk Si and SOI in an eDRAM test chip. It is interesting to note that there is no noticeable difference in the SOI retention time compared to that of the control bulk Si [25].

is undoubtedly a promising beginning of the SOC era driven by patterned SIMOX.

ACKNOWLEDGEMENTS

The author gratefully acknowledges useful discussions with the following colleagues at IBM: Ghavam Shahidi, Paul Agnello, Joel de Souza, Harry Hovel, Siegfried Maurer, Junedong Lee, Herbert Ho, Sundar Iyer, Ken Giewont, Mike Steigerwalt, Tony Domencucci and Dom Schepis.

REFERENCES

[1] J. Margail et al. [in *Silicon-on-Insulator Technology and Devices* Eds. K. Izumi, S. Cristoloveanu, P.L.F. Hemment, G.W. Cullen (Electrochem Soc, 1992) vol.92-13 p.407]

[2] M. Bruel [*Nucl. Instrum. Methods Phys. Res. B (Netherlands)* vol.108 (1996) p.313–9]

[3] G. Shahidi et al. [*IEEE Int. SOI Conf.* Rohnert Park, CA, 1999, p.1–4]

[4] G.K. Celler et al. [*Appl. Phys. Lett. (USA)* vol.48 (1986) p.532]

[5] D.K. Sadana, J.P. de Souza [unpublished]

[6] K. Izumi et al. [*Electrochem. Soc. Proc. (USA)* vol.99-3 (1999) p.225–30]

[7] S. Nakashima, K. Izumi [*J. Mater. Res. (USA)* vol.8 (1993) p.523]

[8] S. Nakashima et al. [*Proc. IEEE SOI Conf.* (1994) p.71]

[9] Y. Li et al. [in *Silicon-on-Insulator Technology and Devices* Eds. K. Izumi, S. Cristoloveanu, P.L.F. Hemment, G.W. Cullen (Electrochem Soc, 1992) vol.92-13, p.368]

[10] K.J. Reeson et al. [*Microelectron. Eng. (Netherlands)* vol.8 (1988) p.163]

[11] M.A. Guerra [*Mater. Sci. Eng. B (Switzerland)* vol.12 (1992) p.145]

[12] S. Bagchi et al. [in *Proc. Int. Symp. on SOI Technology and Devices* Ed. P.L.F. Hemment (Electrochem. Soc, 1996) vol.96]

[13] M.I. Current et al. [in *Ion Implantation Science and Technology* Ed. J.F. Ziegler (1996) p.92–174]

[14] M. Anc, D.K. Sadana [in *Properties of Crystalline Silicon* Ed. R. Hull (INSPEC/IEE, 1999) p.979–91]

[15] S. Nakashima [private communication, 2000]

[16] O.W. Holland et al. [*Appl. Phys. Lett. (USA)* vol.69 (1996) p.574]

[17] D.K. Sadana et al. [*Proc. IEEE SOI Conf.* (1994) p.111]

[18] D.K. Sadana [in *Silicon-on-Insulator Technology and Devices VII* Eds. S. Cristoloveanu, P.L.F. Hemment, K. Izumi, S. Wilson (Electrochem Proc, 1996) vol.96-3, p.3]

[19] S. Cristoloveanu, S. Li [*Electrical Characterization of Silicon-on-Insulator Materials Devices* (Kluwer Academic Publishers, 1995)]

[20] S.T. Liu et al. [*1990 IEEE SOS/SOI Technology Conf. Proc.* p.61]

[21] H.J. Hovel [*Proc. IEEE SOI Conf.* (1997) p.180]

[22] J. Blake et al. [*Proc. IIT 2002* Taos, New Mexico]

[23] K. Tokiguchi et al. [*Proc. IIT 2000* Alpbach, Austria, p.372]

[24] S.K. Iyer et al. [*Proc. VLSI 2000* Hawaii]

[25] H. Ho et al. [*Int. Electron Devices Meet. Tech. Dig. (USA)* (2001)]

Subject Index